T0306006

Contemporary Perspectives in Differential Geometry and its Related Fields

Contemporary Perspectives in Differential Geometry and its Related Fields

Proceedings of the 5th International Colloquium on Differential Geometry and its Related Fields

Veliko Tarnovo, Bulgaria   6 – 10 September 2016

editors

Toshiaki Adachi
Nagoya Institute of Technology, Japan

Hideya Hashimoto
Meijo University, Japan

Milen J Hristov
St Cyril and St Methodius University of Veliko Tarnovo, Bulgaria

World Scientific

EW JERSEY · LONDON · SINGAPORE · BEIJING · SHANGHAI · HONG KONG · TAIPEI · CHENNAI · TOKYO

Published by

World Scientific Publishing Co. Pte. Ltd.

5 Toh Tuck Link, Singapore 596224

USA office: 27 Warren Street, Suite 401-402, Hackensack, NJ 07601

UK office: 57 Shelton Street, Covent Garden, London WC2H 9HE

British Library Cataloguing-in-Publication Data
A catalogue record for this book is available from the British Library.

Cover image:
    (front)   Ovech Fortress, Provadia
    (back)   The way to the fortress, Provadia
Photographed by Toshiaki ADACHI

**CONTEMPORARY PERSPECTIVES IN DIFFERENTIAL GEOMETRY AND
ITS RELATED FIELDS**
**Proceedings of the 5th International Colloquium on Differential Geometry
and its Related Fields**

ISBN 978-981-3220-90-4

Desk Editor: Lai Fun Kwong
Cover Designer: Chin Choon Ng

Printed in Singapore

PREFACE

The *5th International Colloquium on Differential Geometry and its Related Fields* (ICDG2016) was held on the campus of St. Cyril and St. Methodius University of Veliko Tarnovo, Bulgaria, during the period of 6–10 September, 2016. Veliko Tarnovo is a historical place in Bulgaria because it was a capital of the Second Bulgarian Empire, and it is also famous for orthodox temples. ICDG is not a big conference, but as it is not big, participants can discuss quite freely in a friendly environment. We have newcomers from Romania and also from Japan. In this cultural town, participants had many fruitful discussions and exchanges of their ideas.

This volume contains contributions by the main participants in the meeting. These cover contemporary developments on Einstein metrics on Lie groups, geodesics and isometric actions on symmetric spaces, Cartan embeddings of these spaces, Hodge structures on Kähler manifolds, curves under influence of magnetic fields, and so on. We hope this volume will also provide a broad overview of differential geometry and its relationship to other fields in mathematics and physics.

We would like to thank all participants for their contribution in the Colloquium and to this proceedings. We acknowledge also the scientific reviewers who read the articles carefully and gave many important suggestions to the authors.

Enough thanks cannot be expressed to Professors Paskal Piperkov and Milen Hristov for their preparation of this meeting and their hospitality. With their assistance, the participants spent a nice day on an excursion trip to Pleven and Lovech. We here note that Professor Piperkov has left the University of Veliko Tarnovo and became an academician of the Bulgarian Academy of Science. We are grateful to Professor Georgi Ganchev for his advice in our meeting and to Professors Simeon Zamkovoy and Gergana Eneva for their support. We also appreciate the local staff in the University of Veliko Tarnovo for their technical support and hospitality. Lastly, we would like to thank the publishing editor Ms. Lai Fun Kwong and the cover designer for their help.

The Editors
20 April, 2017

The 5th International Colloquium
on Differential Geometry and its Related Fields

6–10 September, 2016 – Veliko Tarnovo, Bulgaria

ORGANIZING COMMITTEE

T. Adachi — Nagoya Institute of Technology, Nagoya, Japan
S. H. Bouyuklieva — St. Cyril and St. Methodius University of
Veliko Tarnovo, Veliko Tarnovo, Bulgaria
H. Hashimoto — Meijo University, Nagoya, Japan
M. J. Hristov — St. Cyril and St. Methodius University of
Veliko Tarnovo, Veliko Tarnovo, Bulgaria

SCIENTIFIC ADVISORY COMMITTEE

G. Ganchev — Bulgarian Academy of Sciences, Sofia, Bulgaria
K. Sekigawa — Niigata University, Niigata, Japan
T. Sunada — Meiji University, Tokyo, Japan

PRESENTATIONS

at Hall 400, Building 5,
St. Cyril and St. Methodius University

1. **Stefka Bouyuklieva** (Univ. Veliko Tarnovo),
   Self-dual codes and their automorphisms

2. **Yusuke Sakane** (Osaka Univ.),
   New homogeneous Einstein metrics on $SU(N)$

3. **Hiroshi Matsuzoe** (Nagoya Inst. Tech.),
   Construction of model selection criterion for q-exponential family

4. **Viktoria Bencheva** (Univ. Veliko Tarnovo),
   Curvature of a barycentric curve and applications to trajectories of rational Bezier

5. **Milen Hristov** (Univ. Veliko Tarnovo),
   Barycentric like structures associated with almost quaternion and almost contact 3-structures

6. **Galya Nakova** (Univ. Veliko Tarnovo),
   Slant null curves on normal almost contact B-metric 3-manifolds with parallel Reeb vector field

7. **Cornelia-Livia Bejan** (Tech. Univ. Iasi),
   New development in differential geometry
   — F-geodesics and some differential operators
   on natural Riemannian extensions —

8. **Osamu Ikawa** (Kyoto Inst. Tech.),
   The geometry of orbits of Hermann type actions

9. **Georgi Ganchev** (Bulgarian Academy Sci.),
   Minimal surfaces in a four-dimensional Euclidean space

10. **Kazuhiro Suzuki** (Nagoya Inst. Tech.),
    Geometrical properties of totally geodesic surfaces in symmetric spaces of Type A

11. **Toshiaki Adachi** (Nagoya Inst. Tech.),
    Trajectory-harps and horns for Kähler magnetic fields

12. **Misa Ohashi** (Nagoya Inst. Tech.),
    Geometrical relations between Lagrangian submanifolds of $G_2/SU(3)$ and totally geodesic surfaces of $G_2/SO(4)$

13. **Nikolay Ivanov** (Univ. Veliko Tarnovo),
    Anomalies in Fermion theories

14. **Hisashi Kasuya** (Tokyo Inst. Tech.),
    Morgan's mixed Hodge structures and geometric structures

15. **Yoshitaka Fukada** (Nagoya Inst. Tech.),
    Realizations of horizontal lift of the Hopf fibration of S^3 to S^2

16. **Takayuki Okuda** (Hiroshima Univ.),
    Totally geodesic submanifolds of Riemmanian symmetric spaces and proper actions of reductive groups on pseudo-Riemannian symmetric spaces

17. **Hideya Hashimoto** (Meijo Univ.),
    Geometrical structures on $G_2/SO(4)$ and related topics

Building 5, St. Cyril and St. Methodius University, September 2016

CONTENTS

HOMOGENEOUS EINSTEIN METRICS
ON COMPLEX STIEFEL MANIFOLDS
AND SPECIAL UNITARY GROUPS

Andreas ARVANITOYEORGOS

Department of Mathematics, University of Patras,
GR-26500 Rion, Greece
E-mail: arvanito@math.upatras.gr

Yusuke SAKANE

Department of Pure and Applied Mathematics,
Graduate School of Information Science and Technology,
Osaka University, Suita, Osaka 565-0871, Japan
E-mail: sakane@math.sci.osaka-u.ac.jp

Marina STATHA

Department of Mathematics, University of Patras,
GR-26500 Rion, Greece
E-mail: statha@master.math.upatras.gr

We obtain new invariant Einstein metrics on the Stiefel manifolds $V_m\mathbb{C}^{m+n} = \mathrm{SU}(m+n)/\mathrm{SU}(n)$ and the compact Lie group $\mathrm{SU}(3+n)$ $(n \geq 5)$ which are not naturally reductive. These are $\mathrm{Ad}(\mathrm{S}(\mathrm{SO}(m) \times \mathrm{U}(1) \times \mathrm{U}(n)))$-invariant metrics and $\mathrm{Ad}(\mathrm{S}(\mathrm{SO}(3) \times \mathrm{U}(1) \times \mathrm{U}(n)))$-invariant metrics respectively.

Keywords: Homogeneous space; Einstein metric; complex Stiefel manifold; special unitary group; isotropy representation; naturally reductive metric; Gröbner basis; resultant.

1. Introduction

A Riemannian manifold (M, g) is called Einstein if it has constant Ricci curvature, i.e. $\mathrm{Ric}_g = \lambda \cdot g$ for some $\lambda \in \mathbb{R}$. A detailed expositions on Einstein manifolds can be found in [4] and for invariant Einstein metrics on homogeneous spaces in [12] and [13]. Even for compact homogeneous spaces, general existence results are difficult to obtain and some methods are described in [5], [6] and also existence and non existence problem is investigated in [14].

Invariant Einstein metrics on the real Stiefel manifolds $\mathrm{SO}(n)/\mathrm{SO}(n-k)$ and the quaternionic Stiefel manifolds $\mathrm{Sp}(n)/\mathrm{Sp}(n-k)$ have been studied through several years by various authors. For a complete review we refer to our work [3], where we proved existence of new invariant Einstein metrics on the Stiefel manifolds $\mathrm{SO}(n)/\mathrm{SO}(n-4)$ and $\mathrm{SO}(7)/\mathrm{SO}(2)$ (in addition to G. Jensen's metrics found in [9]). The first invariant Einstein metrics on the complex Stiefel manifolds $\mathrm{SU}(n)/\mathrm{SU}(n-k)$ were obtain by Jensen in [9], which are $\mathrm{Ad}(\mathrm{S}(\mathrm{U}(n-k) \times \mathrm{U}(k)))$-invariant. To our knowledge there are no other results in the literature.

Left-invariant metrics Einstein metrics on the compact unitary group $\mathrm{SU}(n)$ have been studied in [7] by J.E. D' Atri and W. Ziller who proved existence of several naturally reductive Einstein metrics, and by K. Mori in [10] who proved existence of non naturally reductive Einstein metrics on $\mathrm{SU}(n)$, for $n \geq 6$. Some naturally reducive Einstein metrics on $\mathrm{SU}(3)$ have also been found by G.W. Gibbons, H. Lü and C.N. Pope in [8].

In the present article we investigate invariant Einstein metrics on the complex Stiefel manifolds $\mathrm{SU}(n)/\mathrm{SU}(n-k)$ of orthonormal k-frames in $\mathbb{C}^n$ and left-invariant Einstein metrics on the compact unitary groups $\mathrm{SU}(n)$, which are not naturally reductive. Our main results are the following:

Theorem A. *Let* $m \geq 6$ *and* $n \geq m/2$ *or* $m = 3, 4, 5$ *and* $n \geq 3$. *Then the complex Stiefel manifolds* $\mathrm{SU}(m+n)/\mathrm{SU}(n)$ *admit four invariant Einstein metrics which are* $\mathrm{Ad}\big(\mathrm{S}(\mathrm{SO}(m) \times \mathrm{U}(1) \times \mathrm{U}(n))\big)$*-invariant. Two of them are Jensen's Einstein metrics and the other two metrics are new Einstein metrics.*

Theorem B. *The compact Lie groups* $\mathrm{SU}(3+n)$ *(*$n \geq 5$*) admit non naturally reductive left-invariant and* $\mathrm{Ad}\big(\mathrm{S}(\mathrm{SO}(3) \times \mathrm{U}(1) \times \mathrm{U}(n))\big)$*-invariant Einstein metrics.*

Further information and more detailed statement are given in Theorem 8.1.

2. The Ricci tensor for reductive homogeneous spaces

Let G be a compact semisimple Lie group, K a connected closed subgroup of G and let $\mathfrak{g}$ and $\mathfrak{k}$ be the corresponding Lie algebras. The Killing form B of $\mathfrak{g}$ is negative definite, so we can define an $\mathrm{Ad}(G)$-invariant inner product $-B$ on $\mathfrak{g}$. Let $\mathfrak{g} = \mathfrak{k} \oplus \mathfrak{m}$ be a reductive decomposition of $\mathfrak{g}$ with respect to $-B$ so that $[\,\mathfrak{k}, \mathfrak{m}\,] \subset \mathfrak{m}$ and $\mathfrak{m} \cong T_o(G/K)$.

We recall an expression for the Ricci tensor for a G-invariant Riemannian metric on a reductive homogeneous space G/K. Any G-invariant metric g on G/K is determined by an $\mathrm{Ad}(K)$-invariant scalar product $\langle\ ,\ \rangle$

on $\mathfrak{m}$. Let $\{X_j\}$ be a $\langle\ ,\ \rangle$-orthonormal basis of $\mathfrak{m}$. Then the Ricci tensor r of the metric g is given as follows ([4] p. 381):

$$r(X,Y) = -\frac{1}{2}\sum_i \langle [X, X_i], [Y, X_i]\rangle - \frac{1}{2}B(X,Y)$$

$$+\frac{1}{4}\sum_{i,j} \langle [X_i, X_j], X\rangle\langle [X_i, X_j], Y\rangle. \tag{1}$$

In general, from the equation (1), it is not so easy to obtain the condition for a G-invariant metric being Einstein. We consider a subset of all invariant metrics on G/K. Let H be a closed subgroup of G with $K \subset H \subset G$ and suppose that there is a closed subgroup L of G such that the Lie algebra $\mathfrak{h}$ of H is a direct sum of $\mathfrak{k}$ and the Lie algebra $\mathfrak{l}$ of L: $\mathfrak{h} = \mathfrak{k} \oplus \mathfrak{l}$. According to Arvanitoyeorgos, Dzhepko and Nikonorov [1], [2] we consider invariant metrics on G/K corresponding to $\mathrm{Ad}(H)$-invariant inner products on $\mathfrak{m}$.

We assume that the space $\mathfrak{m}$ can be decomposed into mutually non equivalent irreducible $\mathrm{Ad}(H)$-modules as follows:

$$\mathfrak{m} = \mathfrak{m}_1 \oplus \cdots \oplus \mathfrak{m}_q. \tag{2}$$

Then a G-invariant metric on G/K corresponding to $\mathrm{Ad}(H)$-invariant inner products on $\mathfrak{m}$ can be expressed as

$$\langle\ ,\ \rangle = x_1(-B)|_{\mathfrak{m}_1} + \cdots + x_q(-B)|_{\mathfrak{m}_q}, \tag{3}$$

for positive real numbers $(x_1, \ldots, x_q) \in \mathbb{R}^q_+$. Moreover, the Ricci tensor r for the metric (3) is of the same form

$$r = z_1(-B)|_{\mathfrak{m}_1} + \cdots + z_q(-B)|_{\mathfrak{m}_q},$$

for some real numbers $z_1, \ldots, z_q$, since Ricci tensor r is also $\mathrm{Ad}(H)$-invariant.

Let $\{e_\alpha\}$ be a $(-B)$-orthonormal basis adapted to the decomposition of $\mathfrak{m}$, i.e. $e_\alpha \in \mathfrak{m}_i$ for some i, and $\alpha < \beta$ if $i < j$. We put $A^\gamma_{\alpha\beta} = -B\left([e_\alpha, e_\beta], e_\gamma\right)$ so that $[e_\alpha, e_\beta] = \sum_\gamma A^\gamma_{\alpha\beta} e_\gamma$ and set $\begin{bmatrix} k \\ ij \end{bmatrix} = \sum (A^\gamma_{\alpha\beta})^2$, where the sum is taken over all indices α, β, γ with $e_\alpha \in \mathfrak{m}_i$, $e_\beta \in \mathfrak{m}_j$, $e_\gamma \in \mathfrak{m}_k$ (cf. [14]). Then the positive numbers $\begin{bmatrix} k \\ ij \end{bmatrix}$ are independent of the $(-B)$-orthonormal bases chosen for $\mathfrak{m}_i, \mathfrak{m}_j, \mathfrak{m}_k$, and $\begin{bmatrix} k \\ ij \end{bmatrix} = \begin{bmatrix} k \\ ji \end{bmatrix} = \begin{bmatrix} j \\ ki \end{bmatrix}$.

Let $d_k = \dim \mathfrak{m}_k$. Then, by the same argument as in [11], we have the following:

Lemma 2.1. *The components $r_1, \ldots, r_q$ of the Ricci tensor r of the metric $\langle \, , \, \rangle$ of the form (3) on G/K are given by*

$$r_k = \frac{1}{2x_k} + \frac{1}{4d_k} \sum_{j,i} \frac{x_k}{x_j x_i} \begin{bmatrix} k \\ ji \end{bmatrix} - \frac{1}{2d_k} \sum_{j,i} \frac{x_j}{x_k x_i} \begin{bmatrix} j \\ ki \end{bmatrix} \quad (k = 1, \ldots, q), \quad (4)$$

where the sum is taken over $i, j = 1, \ldots, q$.

By Lemma 2.1, it follows that G-invariant Einstein metrics on $M = G/K$ corresponding to $\mathrm{Ad}(H)$-invariant inner products on $\mathfrak{m}$ are exactly the positive real solutions $(x_1, \ldots, x_q) \in \mathbb{R}_+^q$ of the system of equations $\{r_1 = \lambda, r_2 = \lambda, \ldots, r_q = \lambda\}$, where $\lambda \in \mathbb{R}_+$ is the Einstein constant.

3. A decomposition of $\mathfrak{su}(m + n)$

We consider the homogeneous space $G/H = \mathrm{SU}(m+n)/\mathrm{S}(\mathrm{SO}(m) \times \mathrm{U}(1) \times \mathrm{U}(n))$, where $H = \mathrm{S}(\mathrm{SO}(m) \times \mathrm{U}(1) \times \mathrm{U}(n))$ is a closed subgroup $\mathrm{S}(\mathrm{U}(m) \times \mathrm{U}(n))$ and $\mathrm{SO}(m)$ is the natural subgroup of $\mathrm{SU}(m)$. Then the tangent space $\mathfrak{m}$ of G/H at eH decomposes into two non equivalent irreducible $\mathrm{Ad}(H)$-submodules

$$\mathfrak{m} = \mathfrak{m}_1 \oplus \mathfrak{m}_2,$$

given by

$$\mathfrak{m}_1 = \left\{ \begin{pmatrix} 0 & A \\ -{}^t\bar{A} & 0 \end{pmatrix} : A \in M(m, n; \mathbb{C}) \right\},$$

$$\mathfrak{m}_2 = \left\{ \begin{pmatrix} \sqrt{-1}C & 0 \\ 0 & 0 \end{pmatrix} : C \in \mathrm{Sym}(m), \ \mathrm{tr}\, C = 0 \right\},$$

where $M(m, n; \mathbb{C})$ denotes the set of all $m \times n$ complex matrices and $\mathrm{Sym}(m)$ denotes the set of all $m \times m$ real symmetirc matrices. In fact, $\mathfrak{m}$ is given by $\mathfrak{h}^\perp$ in $\mathfrak{g} = \mathfrak{su}(m+n)$ with respect to $-B$, where B is Killing form of $\mathfrak{su}(m+n)$.

Let $\mathfrak{h} = \mathfrak{h}_0 \oplus \mathfrak{h}_1 \oplus \mathfrak{h}_2$ be the decomposition of $\mathfrak{h}$, the Lie algebra of H, into its center $\mathfrak{h}_0$, and simple ideals as follows:

$$\mathfrak{h}_0 = \left\{ \sqrt{-1} \begin{pmatrix} (b/m)I_m & 0 \\ 0 & (c/n)I_n \end{pmatrix} : b + c = 0, b, c \in \mathbb{R} \right\},$$

$$\mathfrak{h}_1 = \left\{ \begin{pmatrix} A_1 & 0 \\ 0 & 0 \end{pmatrix} : A_1 \in \mathfrak{so}(m) \right\},$$

$$\mathfrak{h}_2 = \left\{ \begin{pmatrix} 0 & 0 \\ 0 & A_2 \end{pmatrix} : A_2 \in \mathfrak{su}(n) \right\}.$$

The Lie algebra $\mathfrak{g}$ splits into $\mathfrak{h}$ and two $\mathrm{Ad}(H)$-irreducible modules as follows:

$$\mathfrak{g} = \mathfrak{h} \oplus \mathfrak{m} = \mathfrak{h}_0 \oplus \mathfrak{h}_1 \oplus \mathfrak{h}_2 \oplus \mathfrak{m}_1 \oplus \mathfrak{m}_2 = \mathfrak{n}_0 \oplus \mathfrak{n}_1 \oplus \mathfrak{n}_2 \oplus \mathfrak{n}_3 \oplus \mathfrak{n}_4. \quad (5)$$

This is an orthogonal decomposition with respect to $-B$. Note that $d_0 = \dim \mathfrak{n}_0 = 1$, $d_1 = \dim \mathfrak{n}_1 = \dfrac{(m-1)m}{2}$, $d_2 = \dim \mathfrak{n}_2 = n^2 - 1$, $d_3 = \dim \mathfrak{n}_3 = 2mn$ and $d_4 = \dim \mathfrak{n}_4 = \dfrac{(m-1)(m+2)}{2}$.

Lemma 3.1. *The submodules in the decomposition* (5) *satisfy the following bracket relations:*

$$[\mathfrak{n}_0, \mathfrak{n}_3] \subset \mathfrak{n}_3, \quad [\mathfrak{n}_0, \mathfrak{n}_j] = (0), (j = 0,1,2,4), \quad [\mathfrak{n}_1, \mathfrak{n}_1] = \mathfrak{n}_1,$$
$$[\mathfrak{n}_1, \mathfrak{n}_2] = (0), \quad [\mathfrak{n}_1, \mathfrak{n}_3] \subset \mathfrak{n}_3, \quad [\mathfrak{n}_1, \mathfrak{n}_4] \subset \mathfrak{n}_4,$$
$$[\mathfrak{n}_2, \mathfrak{n}_2] = \mathfrak{n}_2, \quad [\mathfrak{n}_2, \mathfrak{n}_3] \subset \mathfrak{n}_3, \quad [\mathfrak{n}_2, \mathfrak{n}_4] = (0),$$
$$[\mathfrak{n}_3, \mathfrak{n}_4] \subset \mathfrak{n}_3, \quad [\mathfrak{n}_3, \mathfrak{n}_3] \subset \mathfrak{n}_0 + \mathfrak{n}_1 + \mathfrak{n}_2 + \mathfrak{n}_4, \quad [\mathfrak{n}_3, \mathfrak{n}_3] \subset \mathfrak{n}_1.$$

Therefore, we see that the only non zero numbers (up to permutation of indices) are

$$\begin{bmatrix} 3 \\ 03 \end{bmatrix}, \begin{bmatrix} 1 \\ 11 \end{bmatrix}, \begin{bmatrix} 3 \\ 13 \end{bmatrix}, \begin{bmatrix} 4 \\ 14 \end{bmatrix}, \begin{bmatrix} 2 \\ 22 \end{bmatrix}, \begin{bmatrix} 3 \\ 23 \end{bmatrix}, \begin{bmatrix} 3 \\ 34 \end{bmatrix}. \quad (6)$$

In order to compute the triplets $\begin{bmatrix} i \\ jk \end{bmatrix}$ we need the following lemma, which is a variation of Lemma 5.2 in [3].

Lemma 3.2. *Let $\mathfrak{q}$ be a simple subalgebra of $\mathfrak{g} = \mathfrak{su}(N)$. Consider an orthonormal basis $\{f_j\}$ of $\mathfrak{q}$ with respect to $-B$ (negative of the Killing form of $\mathfrak{su}(N)$), and denote by $B_\mathfrak{q}$ the Killing form of $\mathfrak{q}$. Then, for $i = 1, \ldots, \dim \mathfrak{q}$, we have*

$$\sum_{j,k=1}^{\dim \mathfrak{q}} \left(-B([f_i, f_j], f_k) \right)^2 = \alpha^{\mathfrak{q}}_{\mathfrak{su}(N)},$$

where $\alpha^{\mathfrak{q}}_{\mathfrak{su}(N)}$ is the constant determined by $B_\mathfrak{q} = \alpha^{\mathfrak{q}}_{\mathfrak{su}(N)} \cdot B|_\mathfrak{q}$.

Lemma 3.3. *The triplets $\begin{bmatrix} i \\ jk \end{bmatrix}$ are given by:*

$$\begin{bmatrix} 3 \\ 03 \end{bmatrix} = 1, \quad \begin{bmatrix} 1 \\ 11 \end{bmatrix} = \frac{(m-2)(m-1)m}{4(m+n)}, \quad \begin{bmatrix} 3 \\ 13 \end{bmatrix} = \frac{m(m-1)n}{2(m+n)},$$

$$\begin{bmatrix} 4 \\ 14 \end{bmatrix} = \frac{(m+2)(m-1)m}{4(m+n)}, \quad \begin{bmatrix} 2 \\ 22 \end{bmatrix} = \frac{(n^2-1)n}{m+n}, \quad \begin{bmatrix} 3 \\ 23 \end{bmatrix} = \frac{m(n^2-1)}{m+n},$$

$$\begin{bmatrix} 3 \\ 34 \end{bmatrix} = \frac{(m+2)(m-1)n}{2(m+n)}.$$

Proof. First note that we have $\displaystyle\sum_{j,k=1}^{\dim \mathfrak{su}(m+n)} (-B([f_i, f_j], f_k))^2 = 1$ for $f_i \in$

$\mathfrak{su}(m+n)$. In particular, we see that $\displaystyle\sum_{j,k=1}^{\dim \mathfrak{su}(m+n)} (-B([f_i, f_j], f_k))^2 = 1$ for

$f_i \in \mathfrak{n}_0$. From Lemma 3.1, we see that

$$\begin{bmatrix} 3 \\ 03 \end{bmatrix} = \sum_{j,k=1}^{\dim \mathfrak{su}(m+n)} (-B([f_i, f_j], f_k))^2 = 1.$$

For $f_i \in \mathfrak{su}(n)$, we have

$$\sum_{j,k=1}^{\dim \mathfrak{su}(n)} (-B([f_i, f_j], f_k))^2 = \alpha_{\mathfrak{su}(m+n)}^{\mathfrak{su}(n)} = \frac{n}{m+n}.$$

Note that $\begin{bmatrix} 2 \\ 22 \end{bmatrix} = \displaystyle\sum_{i=1}^{\dim \mathfrak{su}(n)} \sum_{j,k=1}^{\dim \mathfrak{su}(n)} (-B([f_i, f_j], f_k))^2.$ Thus we obtain

$\begin{bmatrix} 2 \\ 22 \end{bmatrix} = \frac{(n^2-1)n}{m+n}$. Similarly, for $f_i \in \mathfrak{su}(m)$, we have

$$\sum_{j,k=1}^{\dim \mathfrak{su}(m)} (-B([f_i, f_j], f_k))^2 = \alpha_{\mathfrak{su}(m+n)}^{\mathfrak{su}(m)} = \frac{m}{m+n}.$$

Note that $\mathfrak{su}(m) = \mathfrak{so}(m) \oplus \mathfrak{m}_2 = \mathfrak{n}_1 + \mathfrak{n}_4$, $[\mathfrak{n}_1, \mathfrak{n}_4] \subset \mathfrak{n}_4$ and $[\mathfrak{n}_4, \mathfrak{n}_4] \subset \mathfrak{n}_1$. Thus we see that

$$\sum_{f_i \in \mathfrak{n}_4} \sum_{j,k=1}^{\dim \mathfrak{su}(m)} (-B([f_i, f_j], f_k))^2 = \begin{bmatrix} 4 \\ 41 \end{bmatrix} + \begin{bmatrix} 1 \\ 44 \end{bmatrix} = \frac{d_4\, m}{m+n} = \frac{(m-1)(m+2)\, m}{2(m+n)}.$$

Hence, we obtain that $\begin{bmatrix} 4 \\ 14 \end{bmatrix} = \frac{(m+2)(m-1)m}{4(m+n)}$. We also have that

$$\sum_{f_i \in \mathfrak{n}_1} \sum_{j,k=1}^{\dim \mathfrak{su}(m)} (-B([f_i, f_j], f_k))^2 = \begin{bmatrix} 4 \\ 14 \end{bmatrix} + \begin{bmatrix} 1 \\ 11 \end{bmatrix} = \frac{d_1\, m}{m+n} = \frac{(m-1)m^2}{2(m+n)}.$$

Hence, we obtain $\begin{bmatrix} 1 \\ 11 \end{bmatrix} = \frac{(m-2)(m-1)m}{4(m+n)}$.

Now, for $f_i \in \mathfrak{so}(m)$, we have $\displaystyle\sum_{j,k=1}^{\dim \mathfrak{su}(m+n)} (-B([f_i, f_j], f_k))^2 = 1$. But we see that

$$\sum_{f_i \in \mathfrak{so}(m)} \sum_{j,k=1}^{\dim \mathfrak{su}(m+n)} (-B([f_i, f_j], f_k))^2 = \begin{bmatrix} 1 \\ 11 \end{bmatrix} + \begin{bmatrix} 4 \\ 14 \end{bmatrix} + \begin{bmatrix} 3 \\ 13 \end{bmatrix},$$

so $d_1 = \begin{bmatrix} 1 \\ 11 \end{bmatrix} + \begin{bmatrix} 4 \\ 14 \end{bmatrix} + \begin{bmatrix} 3 \\ 13 \end{bmatrix}$. Hence, we obtain

$$\begin{bmatrix} 3 \\ 13 \end{bmatrix} = d_1 - \frac{d_1\, m}{m+n} = \frac{d_1\, n}{m+n} = \frac{m(m-1)n}{2(m+n)}.$$

Similarly, we have $d_2 = \begin{bmatrix} 2 \\ 22 \end{bmatrix} + \begin{bmatrix} 3 \\ 23 \end{bmatrix}$ and hence, we obtain

$$\begin{bmatrix} 3 \\ 23 \end{bmatrix} = d_2 - \frac{d_2\, m}{m+n} = \frac{d_2\, m}{m+n} = \frac{m(n^2-1)}{m+n}.$$

For $f_i \in \mathfrak{n}_3$, we see that $\displaystyle\sum_{j,k=1}^{\dim \mathfrak{su}(m+n)} (-B([f_i, f_j], f_k))^2 = 1$ and thus

$$\sum_{f_i \in \mathfrak{n}_3} \sum_{j,k=1}^{\dim \mathfrak{su}(m+n)} (-B([f_i, f_j], f_k))^2$$

$$= \begin{bmatrix} 3 \\ 31 \end{bmatrix} + \begin{bmatrix} 1 \\ 33 \end{bmatrix} + \begin{bmatrix} 3 \\ 30 \end{bmatrix} + \begin{bmatrix} 0 \\ 33 \end{bmatrix} + \begin{bmatrix} 3 \\ 32 \end{bmatrix} + \begin{bmatrix} 2 \\ 33 \end{bmatrix} + \begin{bmatrix} 3 \\ 34 \end{bmatrix} + \begin{bmatrix} 4 \\ 33 \end{bmatrix}.$$

Therefore, we obtain that

$$\begin{bmatrix} 3 \\ 34 \end{bmatrix} = d_3 - \left(2\begin{bmatrix} 3 \\ 34 \end{bmatrix} + 2\begin{bmatrix} 3 \\ 34 \end{bmatrix} + 2\begin{bmatrix} 3 \\ 34 \end{bmatrix} \right) = \frac{(m+2)(m-1)n}{2(m+n)}. \qquad \square$$

4. The Ricci tensor for the Stiefel manifolds $V_m\mathbb{C}^{m+n}$

We now consider the complex Stiefel manifold $G/K = \mathrm{SU}(m+n)/\mathrm{SU}(n)$, and the $\mathrm{Ad}(H)$-invariant decomposition of its tangent space $\mathfrak{p}$ at eK, given by

$$\mathfrak{p} = \mathfrak{h}_0 \oplus \mathfrak{h}_1 \oplus \mathfrak{m}_1 \oplus \mathfrak{m}_2 = \mathfrak{n}_0 \oplus \mathfrak{n}_1 \oplus \mathfrak{n}_3 \oplus \mathfrak{n}_4. \tag{7}$$

We consider a subset of all G-invariant metrics on G/K determined by the $\mathrm{Ad}(H)$-invariant scalar products on $\mathfrak{p}$ given by

$$\langle\ ,\ \rangle = x_0\, (-B)|_{\mathfrak{n}_0} + x_1\, (-B)|_{\mathfrak{n}_1} + x_3\, (-B)|_{\mathfrak{n}_3} + x_4\, (-B)|_{\mathfrak{n}_4}, \tag{8}$$

where $x_i > 0$ $(i = 0, 1, 3, 4)$.

We use the formula for the Ricci tensor in Lemma 2.1 and triplets (6) to obtain the following:

Lemma 4.1. *The components of the Ricci tensor r for the* $\mathrm{Ad}(H)$*-invariant scalar product $\langle \ , \ \rangle$ on $G/K = \mathrm{SU}(m+n)/\mathrm{SU}(n)$ defined by (8) are given as follows:*

$$r_0 = \frac{1}{2x_0} + \frac{1}{4}\begin{bmatrix} 0 \\ 33 \end{bmatrix}\frac{x_0}{x_3{}^2} - \frac{1}{2}\begin{bmatrix} 3 \\ 03 \end{bmatrix}\frac{1}{x_0},$$

$$r_1 = \frac{1}{4d_1}\left(\begin{bmatrix} 1 \\ 11 \end{bmatrix}\frac{1}{x_1} + \begin{bmatrix} 1 \\ 33 \end{bmatrix}\frac{x_1}{x_3{}^2} + \begin{bmatrix} 1 \\ 44 \end{bmatrix}\frac{x_1}{x_4{}^2}\right),$$

$$r_3 = \frac{1}{2x_3} - \frac{1}{2d_3}\left(\begin{bmatrix} 0 \\ 33 \end{bmatrix}\frac{x_0}{x_3{}^2} + \begin{bmatrix} 1 \\ 33 \end{bmatrix}\frac{x_1}{x_3{}^2} + \begin{bmatrix} 4 \\ 33 \end{bmatrix}\frac{x_4}{x_3{}^2}\right),$$

$$r_4 = \frac{1}{2x_4} + \frac{1}{4d_4}\begin{bmatrix} 4 \\ 33 \end{bmatrix}\frac{x_4}{x_3{}^2} - \frac{1}{2d_4}\left(\begin{bmatrix} 1 \\ 44 \end{bmatrix}\frac{x_1}{x_4{}^2} + \begin{bmatrix} 3 \\ 43 \end{bmatrix}\frac{1}{x_4}\right).$$

Then a direct application of Lemma 3.3 gives the following:

Proposition 4.1. *The components of the Ricci tensor r for the* $\mathrm{Ad}(H)$*-invariant scalar product $\langle \ , \ \rangle$ on $G/K = \mathrm{SU}(m+n)/\mathrm{SU}(n)$ defined by (8) are given as follows:*

$$r_0 = \frac{x_0}{4\,x_3{}^2}, \qquad r_1 = \frac{(m-2)}{8(m+n)}\frac{1}{x_1} + \frac{n}{4(m+n)}\frac{x_1}{x_3{}^2} + \frac{(m+2)}{8(m+n)}\frac{x_1}{x_4{}^2},$$

$$r_3 = \frac{1}{2x_3} - \frac{x_0}{4mnx_3{}^2} - \frac{m-1}{8(m+n)}\frac{x_1}{x_3{}^2} - \frac{(m+2)(m-1)}{8m(m+n)}\frac{x_4}{x_3{}^2},$$

$$r_4 = \frac{m}{2(m+n)}\frac{1}{x_4} + \frac{n}{4(m+n)}\frac{x_4}{x_3{}^2} - \frac{m}{4(m+n)}\frac{x_1}{x_4{}^2}.$$

We now consider the system of equations

$$r_0 = r_1, \ r_0 = r_3, \ r_0 = r_4. \tag{9}$$

Then finding Einstein metrics of the form (8) reduces to finding the positive solutions of the system (9). We normalise our equations by setting $x_3 = 1$ and obtain the following system of equations for the variables x_0, x_1 and x_4:

$$\begin{cases} f_1 = 2x_0x_1x_4{}^2(m+n) - (m+2)x_1{}^2 - (m-2)x_4{}^2 - 2nx_1{}^2x_4{}^2 = 0, \\ f_2 = 2x_0(mn+1)(m+n) + (m-1)mnx_1 \\ \qquad + (m-1)(m+2)nx_4 - 4mn(m+n) = 0, \\ f_3 = x_0x_4{}^2(m+n) + mx_1 - 2mx_4 - nx_4{}^3 = 0. \end{cases} \tag{10}$$

5. New invariant Einstein metrics on $V_m\mathbb{C}^{m+n}$

The main result is the following:

Theorem 5.1. *Let $m \geq 6$ and $n \geq m/2$ or $m = 3, 4, 5$ and $n \geq 3$. Then the complex Stiefel manifolds $V_m\mathbb{C}^{m+n}$ admit four $\mathrm{Ad}(\mathrm{S}(\mathrm{SO}(m) \times \mathrm{U}(1) \times \mathrm{U}(n)))$-invariant Einstein metrics. Two of them are Jensen's Einstein metrics and the other two metrics are new Einstein metrics.*

Proof. If $x_1 = x_4$, then we obtain the system of equations

$$\begin{cases} -2mn(m+n) + (m+n)(1+mn)x_0 + (-1+m)(1+m)nx_4 = 0, \\ -m + (m+n)\,x_0\,x_4 - nx_4{}^2 = 0 \end{cases} \quad (11)$$

from (10). From the second equation we obtain $x_0 = (m + nx_4{}^2)/((m + n)x_4)$, and hence

$$n\,(m+n)\,x_4{}^2 - 2n\,(m+n)\,x_4 + m\,n + 1 = 0.$$

Thus we obtain Jensen's Einstein metrics on the complex Stiefel manifolds $V_m\mathbb{C}^{m+n}$, which are determined by the $\mathrm{Ad}(\mathrm{S}(\mathrm{U}(m) \times \mathrm{U}(n)))$-invariant scalar products.

We now work the case $x_1 \neq x_4$. We consider a polynomial ring $R = \mathbb{Q}[z, x_0, x_1, x_4]$ and an ideal I generated by $\{f_1, f_2, f_3, z\,x_0, x_1\,x_4\,(x_1 - x_4) - 1\}$ to find non zero solutions of equations (10) with $x_1 \neq x_4$. We take a lexicographic order $>$ with $z > x_0 > x_1 > x_4$ for a monomial ordering on R. Then, by the aid of computer, we see that a Gröbner basis for the ideal I contains the polynomials $\{g_4(x_4), h_1(x_1, x_4), h_0(x_0, x_4)\}$ given by

$$\begin{cases} g_4(x_4) = n^2(m + 2n + 1)\,x_4{}^4 - 4n^2(m+n)\,x_4{}^3 \\ \qquad + n(2m^2 + 7mn + m + 2n + 6)x_4{}^2 \\ \qquad - 2(3m+2)(m+n)n\,x_4 + (5m+6)(mn+1), \\ h_1(x_1, x_4) = \left(3m^2 + 4mn - m + 2\right)x_1 - 2n(m + 2n + 1)\,x_4{}^3 \\ \qquad + 8n(m+n)\,x_4{}^2 - \left(m^2 + 8mn - 3m + 10\right)x_4, \\ h_0(x_0, x_4) = (m+n)(mn+1)\left(3m^2 + 4mn - m + 2\right)x_0 \\ \qquad + (m-1)mn^2(m+2n+1)\,x_4{}^3 - 4(m-1)mn^2(m+n)\,x_4{}^2 \\ \qquad + (m-1)n\left(2m^3 + 6m^2n + m^2 + 4mn + 5m + 2\right)x_4 \\ \qquad - 2mn(m+n)\left(3m^2 + 4mn - m + 2\right). \end{cases} \quad (12)$$

Now we consider solutions of the equation $g_4(x_4) = 0$. First note that

$$g_4(0) = 6 + 5m + 6mn + 5m^2n > 0$$

and

$$g_4(2) = 6 + 5m + 24n + 2mn + m^2n + 16n^2 > 0.$$

We also see that

$$g_4(4/3) = 1/81(486 + 405m + 9\left(5m^2 + 22m + 96\right)n - 8(19m - 14)n^2 - 256n^3).$$

We claim that, for $n \geq m/2$ and $m \geq 6$, $g_4(4/3) < 0$. We consider $g_4(4/3)$ as a polynomial of n. By expanding at $n = m/2$, we obtain that

$$\begin{aligned}
g_4(4/3) = &-\frac{256}{81}\left(n - \frac{m}{2}\right)^3 - \left(\frac{536m}{81} - \frac{112}{81}\right)\left(n - \frac{m}{2}\right)^2 \\
&+ \frac{1}{81}\left(-299m^2 + 310m + 864\right)\left(n - \frac{m}{2}\right) \\
&+ \frac{1}{162}\left(-95m^3 + 254m^2 + 1674m + 972\right).
\end{aligned} \tag{13}$$

We also see that

$$-299m^2 + 310m + 864 = -299(m - 3)^2 - 1484(m - 3) - 897$$

and

$$\begin{aligned}
&-95m^3 + 254m^2 + 1674m + 972 \\
&= -95(m - 6)^3 - 1456(m - 6)^2 - 5538(m - 6) - 360.
\end{aligned}$$

Thus, for $m \geq 6$ and $n \geq m/2$, we see that $g_4(4/3) < 0$.

For $m = 3, 4, 5$, we can see that $g_4(4/3) < 0$ as follows: For $m = 3$, we have

$$\begin{aligned}
g_4(4/3) &= \frac{1}{81}(1701 + 1863n - 344n^2 - 256n^3) \\
&= -\frac{1}{81}(2718 + 7113(-3 + n) + 2648(-3 + n)^2 + 256(-3 + n)^3),
\end{aligned}$$

thus we see, for $n \geq 3$, $g_4(4/3) < 0$; For $m = 4$, we have

$$\begin{aligned}
g_4(4/3) &= -\frac{2}{81}(-1053 - 1188n + 248n^2 + 128n^3) \\
&= -\frac{2}{81}(1071 + 3756(-3 + n) + 1400(-3 + n)^2 + 128(-3 + n)^3),
\end{aligned}$$

thus we see, for $n \geq 3$, $g_4(4/3) < 0$; For $m = 5$, we have

$$\begin{aligned}
g_4(4/3) &= \frac{1}{81}(2511 + 2979n - 648n^2 - 256n^3) \\
&= -\frac{1}{81}(1296 + 7821(-3 + n) + 2952(-3 + n)^2 + 256(-3 + n)^3),
\end{aligned}$$

thus we see, for $n \geq 3$, $g_4(4/3) < 0$.

Therefore, in the cases above we obtain two positive solutions $x_4 = \alpha_4, \beta_4$ for $g_4(x_4) = 0$, where $0 < \alpha_4 < 4/3$ and $4/3 < \beta_4 < 2$. By substituting these values $x_4 = \alpha_4, \beta_4$ in the equations $h_1(x_1, x_4) = 0$ and

$h_0(x_0, x_4) = 0$ of (12), we obtain two real values of $x_0 = \alpha_0, \beta_0$ and $x_1 = \alpha_1, \beta_1$ and thus two solutions of the system equations (10). But we have to show that these values for $x_0 = \alpha_0, \beta_0$ and $x_1 = \alpha_1, \beta_1$ are positive.

We take a lexicographic order $>$ with $z > x_4 > x_1 > x_0$ for a monomial ordering on R. Then, by the aid of computer, we see that a Gröbner basis for the ideal I contains a polynomial $g_0(x_0)$ given by

$$
\begin{aligned}
g_0(x_0) = {} & \\
& (m + 2n + 1)(mn + 1)\left(3m^2 + 4mn - m + 2\right)^2 (m + n)^4 x_0{}^4 \\
& -2n(3m^2 + 4mn - m + 2)(3m^4 + 24m^3 n + 2m^3 \\
& +32m^2 n^2 + 8m^2 n + 15m^2 + 32mn + 8m + 4)(m + n)^4 x_0{}^3 \\
& +n(10m^8 + 205m^7 n + 7m^7 + 1141m^6 n^2 + 52m^6 n - 9m^6 \\
& +2472m^5 n^3 + 68m^5 n^2 + 386m^5 n + 18m^5 + 2304m^4 n^4 \\
& +16m^4 n^3 + 1509m^4 n^2 - 94m^4 n + 23m^4 + 768m^3 n^5 \\
& +1928m^3 n^3 - 98m^3 n^2 + 241m^3 n - 131m^3 + 768m^2 n^4 \\
& +32m^2 n^3 + 444m^2 n^2 - 154m^2 n + 78m^2 + 160mn^3 \\
& +8mn^2 + 124mn - 20m + 8n + 24)(m + n)^2 x_0{}^2 \\
& -2n^2\left(m^2 + 2mn + 1\right)\left(37m^5 + 256m^4 n + 36m^4 + 512m^3 n^2\right. \\
& \left. -43m^3 + 256m^2 n^3 - 162m^2 +124m + 8\right)(m + n)^2 x_0 \\
& +n^2\Big(768m^3 n^5 + 768m^2\left(3m^2 + 1\right) n^4 + 8m(309m^4 + 2m^3 \\
& +241m^2 + 4m + 20)n^3 + m(1141m^5 + 68m^4 + 1509m^3 \\
& -98m^2 + 444m + 8)n^2 + (205m^7 + 52m^6 + 386m^5 - 94m^4 \\
& +241m^3 - 154m^2 + 124m + 8)n + (m - 1)^2(10m^6 + 27m^5 \\
& +35m^4 + 61m^3 + 110m^2 + 28m + 24)\Big).
\end{aligned}
$$

(14)

Note that, for $m \geq 1$, the coefficients of the polynomial $g_0(x_0)$ are positive for even degree and negative for odd degree terms and hence, if the equation $g_0(x_0) = 0$ has real solutions, then these are all positive. In particular, $x_0 = \alpha_0, \beta_0$ are positive.

We take a lexicographic order $>$ with $z > x_4 > x_0 > x_1$ for a monomial ordering on R. Then, by the aid of computer, we see that a Gröbner basis for the ideal I contains a polynomial $g_1(x_1)$ given by

$$
\begin{aligned}
g_1(x_1) = {} & n^2\left(3m^2 + 4mn - m + 2\right)^2 x_1{}^4 \\
& -8(3m + 2)n^2(m + n)\left(3m^2 + 4mn - m + 2\right)x_1{}^3 \\
& +n(18m^5 + 163m^4 n + 51m^4 + 314m^3 n^2 + 147m^3 n + 27m^3 \\
& +144m^2 n^3 + 348m^2 n^2 + 90m^2 n - 126m^2 + 192mn^3 + 88mn^2 \\
& +20mn - 76m + 64n^3 + 16n^2 - 88n + 72)x_1{}^2 \\
& -2(m - 2)n(m + n)\left(9m^3 + 38m^2 n + 21m^2 + 48mn + 36m + 8n + 28\right)x_1 \\
& +(m - 2)^2(5m + 6)(m + 2n + 1)(mn + 1).
\end{aligned}
$$

(15)

Note that, for $m \geq 3$, the coefficients of the polynomial $g_1(x_1)$ are positive for even degree and negative for odd degree terms and hence, if the equation $g_1(x_1) = 0$ has real solutions, then these are all positive. In particular, $x_1 = \alpha_1, \beta_1$ are positive.   □

Remark 5.1. For $n = 2$, we see that

$$g_4(x_4) = 4(m+5)x_4{}^4 - 16(m+2)x_4{}^3 + 2\left(2m^2 + 15m + 10\right)x_4{}^2$$
$$- 4(m+2)(3m+2)x_4 + (2m+1)(5m+6).$$

We can show that, for $m \geq 3$, $g_4(x_4) > 0$ for all x_4, and hence, there are no real solutions for $g_4(x_4) = 0$. In this case the only $\mathrm{Ad}(S(\mathrm{SO}(m) \times \mathrm{U}(1) \times \mathrm{U}(2)))$-invariant Einstein metrics on the complex Stiefel manifolds $V_m\mathbb{C}^{m+2}$ are Jensen's Einstein metrics.

6. The compact Lie group $\mathrm{SU}(m+n)$

We now consider the compact Lie group $G = \mathrm{SU}(m+n)$, the decomposition (5) of its Lie algebra and a subset of all left-invariant metrics on G, determined by the $\mathrm{Ad}(H) = \mathrm{Ad}(S(\mathrm{SO}(m) \times \mathrm{U}(1) \times \mathrm{U}(n)))$-invariant scalar products $\langle \ , \ \rangle$ on $\mathfrak{g}$ given by

$$x_0\left(-B\right)|_{\mathfrak{h}_0} + x_1\left(-B\right)|_{\mathfrak{h}_1} + x_2\left(-B\right)|_{\mathfrak{h}_2} + x_3\left(-B\right)|_{\mathfrak{m}_1} + x_4\left(-B\right)|_{\mathfrak{m}_2} \quad (16)$$

where $x_i > 0$ $(i = 0, 1, 2, 3, 4)$.

We use the formula for the Ricci tensor in Lemma 2.1 and triplets (6) to obtain the following:

Lemma 6.1. *The components of the Ricci tensor r for the $\mathrm{Ad}(H)$-invariant scalar product $\langle \ , \ \rangle$ on $G = \mathrm{SU}(m+n)$ defined by (16) are given as follows:*

$$r_0 = \frac{1}{2x_0} + \frac{1}{4}\begin{bmatrix}0\\33\end{bmatrix}\frac{x_0}{x_3{}^2} - \frac{1}{2}\begin{bmatrix}3\\03\end{bmatrix}\frac{1}{x_0},$$

$$r_1 = \frac{1}{2x_1} + \frac{1}{4d_1}\left(\begin{bmatrix}1\\11\end{bmatrix}\frac{1}{x_1} + \begin{bmatrix}1\\33\end{bmatrix}\frac{x_1}{x_3{}^2} + \begin{bmatrix}1\\44\end{bmatrix}\frac{x_1}{x_4{}^2}\right)$$
$$- \frac{1}{2d_1}\left(\begin{bmatrix}1\\11\end{bmatrix}\frac{1}{x_1} + \begin{bmatrix}3\\13\end{bmatrix}\frac{1}{x_1} + \begin{bmatrix}4\\14\end{bmatrix}\frac{1}{x_1}\right),$$

$$r_2 = \frac{1}{2x_2} + \frac{1}{4d_2}\left(\begin{bmatrix}2\\22\end{bmatrix}\frac{1}{x_2} + \begin{bmatrix}2\\33\end{bmatrix}\frac{x_2}{x_3{}^2}\right) - \frac{1}{2d_2}\left(\begin{bmatrix}2\\22\end{bmatrix}\frac{1}{x_2} + \begin{bmatrix}3\\23\end{bmatrix}\frac{1}{x_2}\right),$$

$$r_3 = \frac{1}{2x_3} - \frac{1}{2d_3}\left(\begin{bmatrix}0\\33\end{bmatrix}\frac{x_0}{x_3{}^2} + \begin{bmatrix}1\\33\end{bmatrix}\frac{x_1}{x_3{}^2} + \begin{bmatrix}2\\33\end{bmatrix}\frac{x_2}{x_3{}^2} + \begin{bmatrix}4\\33\end{bmatrix}\frac{x_4}{x_3{}^2}\right),$$

$$r_4 = \frac{1}{2x_4} + \frac{1}{4d_4}\begin{bmatrix}4\\33\end{bmatrix}\frac{x_4}{x_3{}^2} - \frac{1}{2d_4}\left(\begin{bmatrix}1\\44\end{bmatrix}\frac{x_1}{x_4{}^2} + \begin{bmatrix}3\\43\end{bmatrix}\frac{1}{x_4}\right).$$

Then by taking into account Lemma 3.3 we obtain the following:

Proposition 6.1. *The components of the Ricci tensor r for the* Ad(H)-*invariant scalar product $\langle\ ,\ \rangle$ on $G = \mathrm{SU}(m+n)$ defined by (16) are given as follows:*

$$r_0 = \frac{x_0}{4\,x_3{}^2},$$

$$r_1 = \frac{(m-2)}{8(m+n)}\frac{1}{x_1} + \frac{n}{4(m+n)}\frac{x_1}{x_3{}^2} + \frac{(m+2)}{8(m+n)}\frac{x_1}{x_4{}^2},$$

$$r_2 = \frac{n}{4(m+n)}\frac{1}{x_2} + \frac{m}{4(m+n)}\frac{x_2}{x_3{}^2},$$

$$r_3 = \frac{1}{2x_3} - \frac{x_0}{4mnx_3{}^2} - \frac{m-1}{8(m+n)}\frac{x_1}{x_3{}^2} - \frac{(n+1)(n-1)}{4n(m+n)}\frac{x_2}{x_3{}^2}$$

$$\quad - \frac{(m+2)(m-1)}{8m(m+n)}\frac{x_4}{x_3{}^2},$$

$$r_4 = \frac{m}{2(m+n)}\frac{1}{x_4} + \frac{n}{4(m+n)}\frac{x_4}{x_3{}^2} - \frac{m}{4(m+n)}\frac{x_1}{x_4{}^2}.$$

We consider the system of equations

$$r_0 = r_1,\ \ r_0 = r_2,\ \ r_0 = r_3,\ \ r_0 = r_4. \tag{17}$$

Then finding Einstein metrics of the form (16) reduces to finding positive solutions of the system (17).

7. Naturally reductive metrics on the compact Lie group SU$(m+n)$

We recall the basic results of D'Atri and Ziller in [7], where they have studied naturally reductive metrics on compact Lie groups and gave a complete classification in the case of simple Lie groups. Let G be a compact, connected semisimple Lie group, L a closed subgroup of G and let $\mathfrak{g}$ be the Lie algebra of G and $\mathfrak{l}$ the subalgebra corresponding to L. Let $\mathfrak{g} = \mathfrak{l} \oplus \mathfrak{m}$ be a reductive decomposition of $\mathfrak{g}$ with respect to the negative of the Killing form $-B$ so that $[\mathfrak{l}, \mathfrak{m}] \subset \mathfrak{m}$ and $\mathfrak{m} \cong T_o(G/L)$. Let $\mathfrak{l} = \mathfrak{l}_0 \oplus \mathfrak{l}_1 \oplus \cdots \oplus \mathfrak{l}_p$ be a decomposition of $\mathfrak{l}$ into ideals, where $\mathfrak{l}_0$ is the center of $\mathfrak{l}$ and $\mathfrak{l}_i$ $(i = 1, \ldots, p)$ are simple ideals of $\mathfrak{l}$. Let $A_0|_{\mathfrak{l}_0}$ be an arbitrary metric on $\mathfrak{l}_0$.

Theorem 7.1. ([7], Theorem 1, p.9 and Theorem 3, p.14) *Under the no-tations above, a left-invariant metric on G of the form*

$$\langle\ ,\ \rangle = x\cdot(-B)|_{\mathfrak{m}} + A_0|_{\mathfrak{l}_0} + u_1\cdot(-B)|_{\mathfrak{l}_1} + \cdots + u_p\cdot(-B)|_{\mathfrak{l}_p}, \qquad (18)$$

where $x, u_1, \cdots, u_p > 0$, is naturally reductive with respect to $G \times L$, where $G \times L$ acts on G by $(g, l)y = gyl^{-1}$.

Moreover, if a left-invariant metric $\langle\ ,\ \rangle$ on a compact simple Lie group G is naturally reductive, then there is a closed subgroup L of G and the metric $\langle\ ,\ \rangle$ is given by the form (18).

Proposition 7.1. *If a left invariant metric $\langle\ ,\ \rangle$ of the form* (16) *on $\mathrm{SU}(m+n)$ is naturally reductive with respect to $\mathrm{SU}(m+n) \times L$ for some closed subgroup L of $\mathrm{SU}(m+n)$, then one of the following holds:*
1) $x_1 = x_4$, 2) $x_3 = x_4$.
Conversely, if one of the conditions 1), 2) *is satisfied, then the metric $\langle\ ,\ \rangle$ of the form* (16) *is naturally reductive with respect to $\mathrm{SU}(m+n) \times L$ for some closed subgroup L of $\mathrm{SU}(m+n)$.*

Proof. Let $\mathfrak{l}$ be the Lie algebra of L. Then we have either $\mathfrak{l} \subset \mathfrak{h}$ or $\mathfrak{l} \not\subset \mathfrak{h}$. First we consider the case of $\mathfrak{l} \not\subset \mathfrak{h}$. Let $\mathfrak{k}$ be the subalgebra of $\mathfrak{su}(m+n)$ generated by $\mathfrak{l}$ and $\mathfrak{h}$. Since $\mathfrak{su}(m+n) = \mathfrak{h}_0 \oplus \mathfrak{h}_1 \oplus \mathfrak{h}_2 \oplus \mathfrak{m}_1 \oplus \mathfrak{m}_2$ is an irreducible decomposition as $\mathrm{Ad}(H)$-modules, we see that the Lie algebra $\mathfrak{k}$ contains at least one of $\mathfrak{m}_1$, $\mathfrak{m}_2$. We first consider the case that $\mathfrak{k}$ contains $\mathfrak{m}_1$. Since $[\mathfrak{m}_1, \mathfrak{m}_1] = \mathfrak{h}_0 \oplus \mathfrak{h}_1 \oplus \mathfrak{h}_2 \oplus \mathfrak{m}_2$, we see that $\mathfrak{k} = \mathfrak{su}(m+n)$. Thus the metric is bi-invariant. If $\mathfrak{k}$ contains $\mathfrak{m}_2$, then $\mathfrak{k} = \mathfrak{h}_0 \oplus \mathfrak{su}(m) \oplus \mathfrak{su}(n)$ and thus $x_1 = x_4$.

In case of $\mathfrak{l} \subset \mathfrak{h}$, noting that the orthogonal complement $\mathfrak{l}^\perp$ of $\mathfrak{l}$ with respect to $-B$ contains the orthogonal complement $\mathfrak{h}^\perp$ of $\mathfrak{h}$, we see that $x_3 = x_4$ by Theorem 7.1.

The converse is a direct consequence of Theorem 7.1. □

8. New invariant Einstein metrics on $\mathrm{SU}(m+n)$

We normalise our equations by setting $x_3 = 1$ and obtain the following system of equations from (17) for the variables x_0, x_1, x_2 and x_4:

$$\begin{cases} g_1 = 2x_0x_1x_4{}^2(m+n) - (m+2)x_1{}^2 - (m-2)x_4{}^2 - 2nx_1{}^2x_4{}^2 = 0, \\ g_2 = x_0x_2(m+n) - mx_2{}^2 - n = 0, \\ g_3 = 2(mn+1)(m+n)x_0 + 2m(n-1)(n+1)x_2 \\ \qquad + (m-1)mn\,x_1 + (m-1)(m+2)nx_4 - 4mn(m+n) = 0, \\ g_4 = x_0x_4{}^2(m+n) + mx_1 - 2mx_4 - nx_4{}^3 = 0. \end{cases} \qquad (19)$$

By solving the system of equations consisting of the first, third and the fourth of equations of (19), we obtain that

$$\left\{ x_0 = \frac{m + nx_4^2}{x_4(m+n)}, \quad x_1 = x_4, \right.$$
$$\left. x_2 = \frac{-nx_4^2(m+n) + 2nx_4(m+n) - mn - 1}{(n^2 - 1)x_4} \right\}, \tag{20}$$

$$\left\{ x_0 = \frac{2n^2 x_4^4 + (7m+2)nx_4^2 + m(5m+6)}{x_4(m+n)(2nx_4^2 + 3m + 2)}, \right.$$
$$x_1 = \frac{(m-2)x_4}{2nx_4^2 + 3m + 2},$$
$$x_2 = \frac{1}{(n^2-1)x_4(3m + 2nx_4^2 + 2)}\left(-n^2(m+2n+1)x_4^4 \right. \tag{21}$$
$$+4n^2(m+n)x_4^3 - n\left(m(2m+7n+1) + 2n + 6\right)x_4^2$$
$$\left. +2(3m+2)n(m+n)x_4 + (-5m-6)(mn+1) \right) \Big\}.$$

We consider the first case (20). Then, from the second equations of (19), that is, $x_0 x_2(m+n) - mx_2^2 - n = 0$, we obtain that

$$(x_4 - 1)^2 \left(n\left(m^2 + mn + n^2 - 1\right)x_4^2 - 2mn(m+n)x_4 + m(mn+1) \right) = 0.$$

For $x_4 = 1$ we see that $x_0 = 1$, $x_1 = x_4 = 1$ and $x_2 = 1$. For the case of

$$n\left(m^2 + mn + n^2 - 1\right)x_4^2 - 2mn(m+n)x_4 + m(mn+1) = 0,$$

we obtain two positive solutions $x_4 = \alpha_4, \beta_4$:

$$x_4 = \alpha_4 = \frac{mn(m+n) - \sqrt{mn(m^2-1)(n^2-1)}}{n(m^2 + mn + n^2 - 1)},$$
$$x_4 = \beta_4 = \frac{m^2 n + mn^2 + \sqrt{mn(m^2-1)(n^2-1)}}{n(m^2 + mn + n^2 - 1)}, \tag{22}$$

and we also see that $x_2 = \dfrac{n}{m}x_4$. Thus we obtain two solutions of the system equations (19):

$$\left\{ x_1 = x_4 = \alpha_4, x_2 = \frac{n}{m}\alpha_4, x_0 = \frac{m + n\alpha_4}{(m+n)\alpha_4} \right\},$$
$$\left\{ x_1 = x_4 = \beta_4, x_2 = \frac{n}{m}\beta_4, x_0 = \frac{m + n\beta_4}{(m+n)\beta_4} \right\},$$

which are naturally reductive Einstein metrics on $\mathrm{SU}(m+n)$ obtained by D'Atri and Ziller [7] (pp. 52 – 53).

We now consider the second case (21). From $x_1 = \dfrac{(m-2)x_4}{2nx_4{}^2 + 3m + 2}$, we see that it should be $m \geq 3$, in order to obtain an invariant metric. From the second of equations (19), that is, $x_0 x_2 (m+n) - m x_2{}^2 - n = 0$, we obtain the polynomial equation $h_4(x_4) = 0$ of x_4:

$$
\begin{aligned}
h_4(x_4) =\ & n^3(m + 2n + 1)\left(m^2 + 2mn + m + 2n^2 - 2\right)x_4{}^8 \\
& -8n^3(m+n)\left(m^2 + 2mn + m + n^2 - 1\right)x_4{}^7 \\
& +n^2(4m^4 + 38m^3 n + 6m^3 + 71m^2 n^2 + 22m^2 n + 3m^2 + 44mn^3 \\
& +19mn^2 + m + 4n^4 + 8n^3 + 6n^2 - 8n - 10)x_4{}^6 \\
& -4n^2(m+n)(7m^3 + 20m^2 n + 7m^2 + 10mn^2 + 8mn + 4m \\
& +4n^2 - 4)x_4{}^5 + n(4m^5 + 86m^4 n + 4m^4 + 184m^3 n^2 + 76m^3 n \\
& +16m^3 + 117m^2 n^3 + 138m^2 n^2 + 51m^2 n + 12m^2 + 12mn^4 \\
& +84mn^3 + 40mn^2 - 4mn + 8n^4 + 4n^3 + 8n^2 - 4n - 16)x_4{}^4 \qquad (23) \\
& -2n(m+n)(12m^4 + 62m^3 n + 14m^3 + 31m^2 n^2 + 64m^2 n \\
& +29m^2 + 32mn^2 + 8mn + 16m + 4n^2 - 4)x_4{}^3 \\
& +(56m^5 n + 152m^4 n^2 + 82m^4 n + 10m^4 + 106m^3 n^3 + 217m^3 n^2 \\
& +88m^3 n + 17m^3 + 9m^2 n^4 + 152m^2 n^3 + 109m^2 n^2 + 72m^2 n \\
& +10m^2 + 12mn^4 + 40mn^3 + 64mn^2 - 4m + 4n^4 + 4n^2 - 8)x_4{}^2 \\
& -2m(3m + 2)(5m + 6)(m+n)\left(2mn + n^2 + 1\right)x_4 \\
& +m(5m + 6)^2(m+n)(mn + 1) = 0.
\end{aligned}
$$

Now we consider the case $m = 3$. In this case we see that

$$
\begin{aligned}
h_4(x_4) =\ & 4n^3(n + 2)\left(n^2 + 3n + 5\right)x_4{}^8 \\
& - 8n^3(n + 3)\left(n^2 + 6n + 11\right)x_4{}^7 \\
& + 2n^2\left(2n^4 + 70n^3 + 351n^2 + 608n + 253\right)x_4{}^6 \\
& \quad 8n^2(n \mid 3)\left(17n^2 \mid 102n \mid 130\right)x_4{}^5 \\
& + n\left(44n^4 + 1309n^3 + 6338n^2 + 9461n + 1820\right)x_4{}^4 \\
& - 2n(n + 3)\left(379n^2 + 2274n + 1655\right)x_4{}^3 \\
& + \left(121n^4 + 4350n^3 + 19348n^2 + 23274n + 1339\right)x_4{}^2 \\
& - 1386(n + 3)\left(n^2 + 6n + 1\right)x_4 + 1323(n + 3)(3n + 1).
\end{aligned}
$$

Note that the coefficients of even degree of the polynomial $h_4(x_4)$ are positive and odd degree are negative. Thus, if there exists a real solution of $h_4(x_4) = 0$, it is always positive. We claim that there are real solutions of $h_4(x_4) = 0$ for $n \geq 5$. Consider the value of $h_4(x_4)$ at $x_4 = 1$. We see that

$$
\begin{aligned}
h_4(1) =\ & -4n^5 - 38n^4 - 16n^3 + 736n^2 + 2060n + 1150 \\
=\ & -4(n - 5)^5 - 138(n - 5)^4 - 1776(n - 5)^3 - 10204(n - 5)^2 \\
& -23280(n - 5) - 8400 < 0.
\end{aligned}
$$

We also have

$$h_4\left(\frac{6}{5}\right) = \frac{9}{390625}\big(20736n^6 - 48960n^5 - 559484n^4 + 5754780n^3$$
$$+40385725n^2 + 76331250n + 39390625\big)$$
$$= 20736(n-5)^6 + 573120(n-5)^5 + 5992516(n-5)^4$$
$$+34165100(n-5)^3 + 175984825(n-5)^2$$
$$+867855000(n-5) + 1971360000 > 0$$

and

$$h_4(0) = 1323(n+3)(3n+1) > 0.$$

Thus we see that there are at least two positive solutions $x_4 = \alpha_4, \beta_4$ with $0 < \alpha_4 < 1$ and $1 < \beta_4 < 6/5$. To get more solutions we consider the value $h_4(x_4)$ at $x_4 = \frac{63}{11\,n}$. We see that, for $n \geq 6$,

$$h_4\left(\frac{63}{11\,n}\right) = -\frac{2646}{214358881n^5}\big(857435524n^6 + 6666745833n^5$$
$$-12579650424n^4 - 139083355176n^3 - 154477369620n^2$$
$$-1250233749765n - 3751410132540\big)$$
$$= -\frac{2646}{214358881n^5}\big(857435524(n-6)^6 + 37534424697(n-6)^5$$
$$+650437907526(n-6)^4 + 5663154998208(n-6)^3$$
$$+25693535331468(n-6)^2 + 54211242295035(n-6)$$
$$+28685897800182\big) < 0.$$

We also see that the value $h_4(x_4)$ at $x_4 = \frac{139}{11\,n}$ is given by

$$h_4\left(\frac{139}{11\,n}\right) = \frac{2}{214358881n^5}\big(619068448328n^7 + 15285651253268n^6$$
$$-5531951718740599n^5 - 112424896352119912n^4$$
$$+9130623975688348888n^3 + 542136577141727724n^2$$
$$+18308840308826803679n + 2787073344233673620\big)$$
$$= \frac{2}{214358881n^5}\big(619068448328(n-26)^7 + 127956108848964(n-26)^6$$
$$+10619662116100037(n-26)^5 + 452664788402750418(n-26)^4$$
$$+10457179897884737640(n-26)^3 + 124075623506627804284(n-26)^2$$
$$+58918195509984224815(n-26) + 5624153791074508074\big).$$

Thus we see that, for $n \geq 26$, $h_4\left(\frac{139}{11\,n}\right) > 0$ and there are two more

positive solutions $x_4 = \gamma_4, \delta_4$ with $0 < \gamma_4 < \dfrac{63}{11n}$ and $\dfrac{63}{11\,n} < \delta_4 < \dfrac{139}{11\,n}$.
Note that, for $n = 25$, we see that $\dfrac{139}{11\,n} \approx 0.505455$ and $h_4\left(\dfrac{139}{11\,n}\right) =$
$-\dfrac{155767386137975880912}{418669689453125} < 0.$

In fact, for $n = 25$, we have the following figures of graph of $h_4(x_4)$:

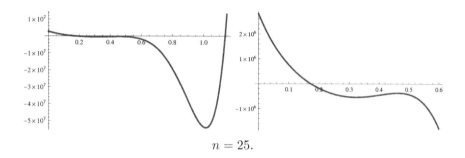

$$n = 25.$$

For $n = 26$, we have the following figures of graph of $h_4(x_4)$:

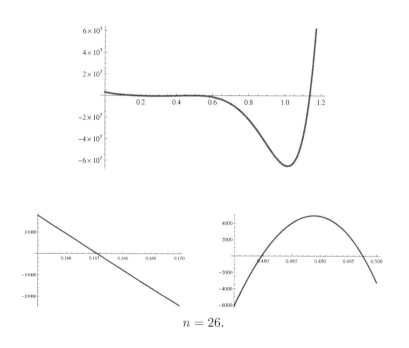

$$n = 26.$$

From (21), we see that the values for x_0 and x_1 corresponding to the solutions of x_4 are positive.

To see that the value for x_2 is positive, we consider the equations $h_4(x_4) = 0$ and the third equation of (21). By substituting $m = 3$ in the third of equations (21), we obtain the equation $p(x_2, x_4) = 0$, where

$$p(x_2, x_4) = 2n^3 x_2 x_4{}^3 + 2n^3 x_4{}^4 - 4n^3 x_4{}^3 + 11n^2 x_2 x_4 + 4n^2 x_4{}^4 - 12n^2 x_4{}^3$$
$$+23n^2 x_4{}^2 - 22n^2 x_4 - 2n x_2 x_4{}^3 + 27n x_4{}^2 - 66n x_4 + 63n - 11 x_2 x_4 + 21.$$

By the aid of computer, we compute the resultant of the polynomials $\{h_4(x_4), p(x_2, x_4)\}$ and obtain a polynomial $h_2(x_2)$ of x_2 given by

$$
\begin{aligned}
h_2(x_2) =\ & 54(n+3)\left(n^2 + 3n + 5\right)\left(2n^2 + 6n + 11\right)^2 x_2{}^8 \\
& -108(n+3)\left(2n^2 + 6n + 11\right)\left(8n^4 + 48n^3 + 124n^2 + 156n + 55\right) x_2{}^7 \\
& +3(2016n^7 + 23808n^6 + 122688n^5 + 360173n^4 + 640074n^3 + 660655n^2 \\
& +334812n + 40170)x_2{}^6 - 12n(n+3)(1008n^5 + 8496n^4 + 29520n^3 \\
& +53885n^2 + 49857n + 17159)x_2{}^5 + 2n(7560n^6 + 82080n^5 + 357423n^4 \\
& +798255n^3 + 951061n^2 + 531765n + 81068)x_2{}^4 - 4n^2(n+3)(3024n^4 \\
& +21456n^3 + 54318n^2 + 58929n + 22198)x_2{}^3 + n^2(6048n^5 + 55296n^4 \\
& +186804n^3 + 281753n^2 + 173678n + 24553)x_2{}^2 \\
& -4n^3(n+3)(6n+13)(72n^2 + 204n + 115)x_2 \\
& +2n^3(n+2)(3n+1)(6n+13)^2.
\end{aligned}
$$

Now the coefficients of even degree of the polynomial of $h_2(x_2)$ of x_2 are positive and odd degree are negative. Thus we see that the real solutions of $h_2(x_2) = 0$ are always positive and hence the values for x_2 given by (21) for $m = 3$ are positive. We also see that the obtained solutions for x_4 have the property that $x_4 \neq 1 = x_3$ and $x_4 \neq x_1$ and hence, the Einstein metrics are not naturally reductive.

Thus we obtain

Theorem 8.1. *The compact Lie groups* $\mathrm{SU}(3+n)$ *(n $\geq$ 5) admit, besides a bi-invariant metric, at least six* $\mathrm{Ad}(\mathrm{S}(\mathrm{SO}(3) \times \mathrm{U}(1) \times \mathrm{U}(n)))$-*invariant Einstein metrics for* $n \geq 26$. *Two of them are naturally reductive Einstein metrics (which are* $\mathrm{Ad}(\mathrm{S}(\mathrm{U}(3) \times \mathrm{U}(n)))$-*invariant) and the other four metrics are non naturally reductive Einstein metrics. For* $5 \leq n \leq 25$, *there are at least four* $\mathrm{Ad}(\mathrm{S}(\mathrm{SO}(3) \times U(1) \times \mathrm{U}(n)))$-*invariant Einstein metrics. Two of them are naturally reductive Einstein metrics (which are* $\mathrm{Ad}(\mathrm{S}(\mathrm{U}(3) \times \mathrm{U}(n)))$-*invariant) and the other two metrics are non naturally reductive Einstein metrics. For* $n = 2, 4$, *there are two naturally reductive Einstein metrics and for* $n = 3$ *there is one naturally reductive Einstein metric (which are* $\mathrm{Ad}(\mathrm{S}(\mathrm{U}(3) \times \mathrm{U}(n)))$-*invariant).*

Acknowledgments

The work was supported by Grant #E.037 from the Research Committee of the University of Patras (Programme K. Karatheodori) and JSPS KAKENHI Grant Number 16K05130.

References

[1] A. Arvanitoyeorgos, V. V. Dzhepko and Yu. G. Nikonorov, Invariant Einstein metrics on some homogeneous spaces of classical Lie groups, *Canad. J. Math.* **61**(6), 1201–1213 (2009).

[2] A. Arvanitoyeorgos, V.V. Dzhepko and Yu. G. Nikonorov, Invariant Einstein metrics on certain Stiefel manifolds, *Differential Geometry and its Applications, Proc. Conf., in Honor of Leonard Euler,* Olomouc, August 2007, World Sci., 35–44 (2008).

[3] A. Arvanitoyeorgos, Y. Sakane and M. Statha, New homogeneous Einstein metrics on Stiefel manifolds, *Differential Geom. Appl.* **35**(S1), 2–18 (2014).

[4] A.L. Besse, *Einstein Manifolds,* Springer-Verlag, Berlin, 1986.

[5] C. Böhm, Homogeneous Einstein metrics and simplicial complexes, *J. Differential Geom.* **67**(1), 79–165 (2004).

[6] C. Böhm, M. Wang and W. Ziller, A variational approach for compact homogeneous Einstein manifolds, *Geom. Func. Anal.* **14** (4), 681–733 (2004).

[7] J. E. D'Atri and W. Ziller, Naturally reductive metrics and Einstein metrics on compact Lie groups, *Memoirs A.M.S.* **19** (215) (1979).

[8] G.W. Gibbons, H. Lü and C.N. Pope, Einstein metrics on group manifolds and cosets, *J. Geom. Physics* **61**, 947–960 (2011).

[9] G. Jensen, Einstein metrics on principal fiber bundles, *J. Differential Geom.* **8**, 599–614 (1973).

[10] K. Mori, Left Invariant Einstein Metrics on $SU(N)$ that are not Naturally Reductive, Master Thesis (in Japanese) Osaka University 1994, English Translation: Osaka University RPM 96010 (preprint series) 1996.

[11] J-S. Park and Y. Sakane, Invariant Einstein metrics on certain homogeneous spaces, *Tokyo J. Math.* **20**(1), 51–61 (1997).

[12] M. Wang, Einstein metrics from symmetry and bundle constructions, in *Surveys in Differential Geometry : Essays on Einstein Manifolds,* Surv. Diff. Geom. VI, Int. Press, Boston, MA 1999.

[13] M. Wang, Einstein metrics from symmetry and bundle constructions: A sequel, in *Differential Geometry :Under the Influence of S.-S. Chern,* Advanced Lectures in Mathematics, vol. 22, 253–309, Higher Education Press/International Press, 2012.

[14] M. Wang and W. Ziller, Existence and non-existence of homogeneous Einstein metrics, *Invent. Math.* **84**, 177–194 (1986).

Received December 16, 2016

Contemporary Perspectives
in Differential Geometry
and its Related Fields   21 – 32

GEODESICS OF RIEMANNIAN SYMMETRIC SPACES
INCLUDED IN REFLECTIVE SUBMANIFOLDS

Takayuki OKUDA

Graduate School of Sciences, Hiroshima University,
1-3-1 Kagamiyama, Higashi-Hiroshima, 739-8526 Japan
E-mail: okudatak@hiroshima-u.ac.jp

Let M be a Riemannian symmetric space of compact type, and we denote by
G the identity component of the isometry group $\mathrm{Isom}(M)$ of M. In this paper,
we give an algorithm to classify G-conjugacy classes of geodesics in a given
reflective submanifold of M in terms of Satake diagrams.

Keywords: Riemannian symmetric spaces of compact type; reflective submanifolds; Satake diagrams.

1. Introduction

Let M be a Riemannian symmetric space of compact, and we denote by G
the identity component of the isometry group $\mathrm{Isom}(M)$ of M. We define
the set of all G-conjugacy classes of maximal geodesics in M by

$$\mathrm{Geod}(M) := \{ [\ell]_G \mid \ell : \mathbb{R} \to M \text{ is a maximal geodesic of } M \},$$

where $[\ell]_G$ denotes the G-conjugacy class of the maximal geodesic ℓ in M.

For a totally geodesic submanifold complete submanifold L of M, any
maximal geodesic of L can be considered as a maximal geodesic of M. We
put

$$\mathrm{Geod}(M; L) := \{ [\ell]_G \mid \ell : \mathbb{R} \to L \text{ is a maximal geodesic of } L \}$$
$$(\subset \mathrm{Geod}(M)).$$

As a subclass of totally geodesic submanifolds in M, Leung [4] intro-
duces the concept of reflective submanifolds as follows: a submanifold L of
M is said to be *reflective* if there exists an involutive isometry τ on M such
that L is a connected component of the set of all fixed points of τ. For
the cases where M is irreducible and simply-connected, Leung [5, 6] give
classifications of all reflective submanifolds.

In this paper, we give an algorithm to classify G-conjugacy classes of
geodesics in a given reflective submanifold of M in terms of Satake diagrams
(see Theorems 3.1 and 3.2).

2. Weighted Dynkin diagrams and Satake diagrams

In this section, we set up our notation for Satake diagrams and weighted Dynkin diagrams.

2.1. Labelled Dynkin diagrams and weighted Dynkin diagrams

Let $\mathfrak{g}$ be a compact semisimple Lie algebra. For the simplicity, fix a connected compact Lie group G with $\mathrm{Lie}\,G \simeq \mathfrak{g}$. In this subsection, we fix our terminology of labelled Dynkin diagram and weighted Dynkin diagrams of $\mathfrak{g}$.

For each maximal abelian subspace $\mathfrak{t}_0$ of $\mathfrak{g}$, the root system of $(\mathfrak{g}, \mathfrak{t}_0)$ is defined as a finite subset of $(\sqrt{-1}\mathfrak{t}_0)^*$. Let us put

$$\Theta_{\mathfrak{g}} := \left\{ (\mathfrak{t}_0, \Pi_0) \; \middle| \; \begin{array}{l} \mathfrak{t}_0 \text{ is a maximal abelian subspace of } \mathfrak{g}, \\ \Pi_0 \text{ is a simple system of the root system of } (\mathfrak{g}, \mathfrak{t}_0) \end{array} \right\}.$$

Then for each $(\mathfrak{t}_0, \Pi_0)$, the Dynkin diagram $D_{\mathfrak{g}}(\Pi_0)$ of $\mathfrak{g}$ with nodes Π_0 is defined.

For any $(\mathfrak{t}_1, \Pi_1), (\mathfrak{t}_2, \Pi_2) \in \Theta_{\mathfrak{g}}$, there uniquely exists a bijection $\phi_{\Pi_1,\Pi_2} : \Pi_1 \to \Pi_2$ such that ϕ_{Π_1,Π_2} is induced by a linear isomorphisms $\mathrm{Ad}(g)|_{\mathfrak{t}_2} : \mathfrak{t}_2 \xrightarrow{\sim} \mathfrak{t}_1$ for some $g \in G$ with $\mathrm{Ad}(g) \cdot \mathfrak{t}_2 = \mathfrak{t}_1$ and $\mathrm{Ad}(g)^*(\Pi_1) = \Pi_2$. Then the bijection ϕ_{Π_1,Π_2} gives an isomorphism between the Dynkin diagrams $D_{\mathfrak{g}}(\Pi_1)$ and $D_{\mathfrak{g}}(\Pi_2)$ of $\mathfrak{g}$.

Let us fix a finite set Π with $\#\Pi = \mathrm{rank}\,\mathfrak{g}$ and a family of bijections $\{\iota_{\Pi_0} : \Pi \xrightarrow{\sim} \Pi_0\}_{(\mathfrak{t}_0,\Pi_0)}$ satisfying that $\phi_{\Pi_1,\Pi_2} \circ \iota_{\Pi_1} = \iota_{\Pi_2}$ for any $(\mathfrak{t}_1, \Pi_1), (\mathfrak{t}_2, \Pi_2) \in \Theta_{\mathfrak{g}}$. Then one can define the diagram $D_{\mathfrak{g}}$ with nodes Π such that $\iota_{\Pi_0} : \Pi \xrightarrow{\sim} \Pi_0$ induces an isomorphism between $D_{\mathfrak{g}}$ and $D_{\mathfrak{g}}(\Pi_0)$ for each $(\mathfrak{t}_0, \Pi_0) \in \Theta_{\mathfrak{g}}$. Throughout this paper, such the diagram $D_{\mathfrak{g}}$ will be called *the Dynkin diagram of $\mathfrak{g}$ labelled by Π*.

In the setting above, we say that a map $\Phi : \Pi \to \mathbb{R}$ is *a weighted Dynkin diagram on $D_{\mathfrak{g}}$*. For each α, the value $\Phi(\alpha)$ is called *the weight of Φ on the node α*.

The set of all weighted Dynkin diagrams on $D_{\mathfrak{g}}$ will be denoted by $\mathrm{Map}(\Pi, \mathbb{R})$. We say that a weighted Dynkin diagram $\Phi : \Pi \to \mathbb{R}$ is *trivial* if $\Phi(\alpha) = 0$ for any $\alpha \in \Pi$, and that it is *non-negative* if $\Phi(\alpha) \geq 0$ for any α. The set of non-trivial and non-negative weighted Dynkin diagrams on $D_{\mathfrak{g}}$ will be denoted by $\mathrm{Map}^*(\Pi, \mathbb{R}_{\geq 0})$.

Example 2.1. Let $\mathfrak{g} = \mathfrak{o}(10)$ (the compact simple Lie algebra of type D_5).

Then

is a Dynkin diagram $D_{o(10)}$ labelled by $\Pi = \{\alpha_1, \ldots, \alpha_5\}$. Then the set of non-trivial and non-negative weighted Dynkin diagrams $\mathrm{Map}^*(\Pi, \mathbb{R}_{\geq 0})$ on $D_{\mathfrak{g}}$ can be written as

$$
\left\{
\begin{array}{c|c}
\text{\raisebox{-1em}{}} &
\begin{array}{l}
a_i \in \mathbb{R}_{\geq 0} \ (i = 1, \ldots, 5), \\
(a_1, \ldots, a_5) \neq (0, \ldots, 0)
\end{array}
\end{array}
\right\}.
$$

We remark that the two weighted Dynkin diagrams Φ_1 and Φ_2 represented as

$$
\Phi_1 = \text{} \quad \text{and} \quad \Phi_2 = \text{}
$$

are different since the weights of these on the nodes α_4 and α_5 are different.

2.2. Satake diagrams

Let $\mathfrak{g}$ be a compact semisimple Lie algebra and $(\mathfrak{g}, \mathfrak{k})$ be a symmetric pair, that is, $\mathfrak{k}$ is a subalgebra of $\mathfrak{g}$ with an involution σ on $\mathfrak{g}$ such that $\mathfrak{k} := \mathfrak{g}^{\sigma} = \{X \in \mathfrak{g} \mid \sigma(X) = X\}$. In this subsection, we recall the definition of the Satake diagram of $(\mathfrak{g}, \mathfrak{k})$ (see [1, 8] for the details of Satake diagrams).

We put $\mathfrak{s} := \mathfrak{g}^{-\sigma} = \{X \in \mathfrak{g} \mid \sigma(X) = -X\}$. Then $\mathfrak{g} = \mathfrak{k} + \mathfrak{s}$ gives a decomposition of $\mathfrak{g}$ as a real vector space.

For a maximal abelian subspace $\mathfrak{a}_0$ of $\mathfrak{s}$, one can define the restricted root system of $(\mathfrak{g}, \mathfrak{a}_0)$ as a finite subset of $(\sqrt{-1}\mathfrak{a}_0)^*$. In this situation, for a maximal abelian subspace $\mathfrak{t}_0$ of $\mathfrak{g}$ with $\mathfrak{a} \subset \mathfrak{t}$, there exists a simple system Π_0 of the root system of $(\mathfrak{g}, \mathfrak{t}_0)$ such that

$$
\Pi_0^- := \{\alpha|_{\sqrt{-1}\mathfrak{a}_0} \in (\sqrt{-1}\mathfrak{a}_0)^* \mid \alpha \in \Pi_0\} \setminus \{0\}
$$

is a simple system of the restricted root system of $(\mathfrak{g}, \mathfrak{a}_0)$. For each triple $(\mathfrak{a}_0, \mathfrak{t}_0, \Pi_0)$ as above, *the Satake diagram* $S_{(\mathfrak{g}, \mathfrak{k})}(\Pi_0)$ consists of the following three data: the Dynkin diagram $D_{\mathfrak{g}}(\Pi_0)$ of $\mathfrak{g}$ with nodes Π_0; black nodes $\Pi_0^b := \{\alpha \in \Pi_0 \mid \alpha|_{\sqrt{-1}\mathfrak{a}_0} = 0\}$ in $S_{(\mathfrak{g}, \mathfrak{k})}(\Pi_0)$; and arrows joining the nodes

α and β in $\Pi_0 \setminus \Pi_0^b$ in $S_{(\mathfrak{g},\mathfrak{k})}(\Pi_0)$ whose restrictions $\alpha|_{\sqrt{-1}\mathfrak{a}_0}$ and $\beta|_{\sqrt{-1}\mathfrak{a}_0}$ are the same.

Throughout this paper, we put

$$\Theta_{(\mathfrak{g},\mathfrak{k})} := \left\{ (\mathfrak{a}_0, \mathfrak{t}_0, \Pi_0) \,\middle|\, \begin{array}{l} \mathfrak{a}_0 \text{ is a maximal abelian subspaces of } \mathfrak{s}; \\ (\mathfrak{t}_0, \Pi_0) \in \Theta_{\mathfrak{g}} \text{ satisfies that } \Pi_0^- \text{ defined above} \\ \text{is a simple system of the restricted root} \\ \text{system of } (\mathfrak{g}, \mathfrak{a}_0) \end{array} \right\}.$$

Then for any $(\mathfrak{a}_1, \mathfrak{t}_1, \Pi_1), (\mathfrak{a}_2, \mathfrak{t}_2, \Pi_2) \in \Theta_{(\mathfrak{g},\mathfrak{k})}$, the bijection $\phi_{\Pi_1,\Pi_2} : \Pi_1 \to \Pi_2$ defined in Section 2.1 induces an isomorphism between the Satake diagrams $S_{(\mathfrak{g},\mathfrak{k})}(\Pi_1)$ and $S_{(\mathfrak{g},\mathfrak{k})}(\Pi_2)$.

We denote by $D_{\mathfrak{g}}$ the Dynkin diagram of $\mathfrak{g}$ labelled by Π (see Section 2.1). Then one can define *the Satake diagram $S_{(\mathfrak{g},\mathfrak{k})}$ labelled by Π* (by defining the set of black nodes in Π and arrows joining some pairs of non-black nodes) such that $\iota_{\Pi_0} : \Pi \xrightarrow{\sim} \Pi_0$ induces an isomorphism between $S_{(\mathfrak{g},\mathfrak{k})}$ and $S_{(\mathfrak{g},\mathfrak{k})}(\Pi_0)$ for each $(\mathfrak{a}_0, \mathfrak{t}_0, \Pi_0) \in \Theta_{(\mathfrak{g},\mathfrak{k})}$.

It should be remarked that the Satake diagram is defined for each G-conjugacy class of symmetric pairs in the following sense. Let $(\mathfrak{g}, \mathfrak{k})$ be a compact semisimple symmetric pair as before and fix any $g \in G$. Then $(\mathfrak{g}, \mathrm{Ad}(g) \cdot \mathfrak{k})$ is also a symmetric pair. Furthermore,

$$\Theta_{(\mathfrak{g},\mathrm{Ad}(g)\cdot\mathfrak{k})} = \left\{ (\mathrm{Ad}(g) \cdot \mathfrak{a}_0, \mathrm{Ad}(g) \cdot \mathfrak{t}_0, \mathrm{Ad}(g^{-1})^* \cdot \Pi_0) \,\middle|\, (\mathfrak{a}_0, \mathfrak{t}_0, \Pi_0) \in \Theta_{(\mathfrak{g},\mathfrak{k})} \right\}$$

where $\mathrm{Ad}(g^{-1})^* : (\sqrt{-1}\mathfrak{t}_0)^* \to (\sqrt{-1}(\mathrm{Ad}(g)\cdot\mathfrak{t}_0))^*$ is the linear map induced by the map $\mathrm{Ad}(g^{-1})|_{\mathrm{Ad}(g)\cdot\mathfrak{t}_0} : \mathrm{Ad}(g) \cdot \mathfrak{t}_0 \to \mathfrak{t}_0$. In particular, the Satake diagram $S_{(\mathfrak{g},\mathfrak{k})}$ of $(\mathfrak{g}, \mathfrak{k})$ labelled by Π defined above is also a Satake diagram of $(\mathfrak{g}, \mathrm{Ad}(g) \cdot \mathfrak{k})$.

Example 2.2. Let $(\mathfrak{g}, \mathfrak{k}) = (\mathfrak{o}(10), \mathfrak{u}(5))$ as a compact symmetric pair of type DIII. Then the Satake diagram $S_{(\mathfrak{o}(10),\mathfrak{u}(5))}$ labelled by $\Pi = \{\alpha_1, \ldots, \alpha_5\}$ can be drawn as

Here, in the Satake diagram $S_{(\mathfrak{o}(10),\mathfrak{u}(5))}$, the set of black nodes is $\{\alpha_1, \alpha_3\}$ and α_4 and α_5 are joined by an arrow.

2.3. Weighted Dynkin diagrams matching Satake diagrams

Let us consider the same setting in Section 2.2. We denote by $D_{\mathfrak{g}}$ and $S_{(\mathfrak{g},\mathfrak{k})}$ the Dynkin diagram and the Satake diagram of $\mathfrak{g}$ and $(\mathfrak{g}, \mathfrak{k})$ labelled by Π, respectively (Sections 2.1 and 2.2 for the notation).

Definition 2.1 ([7, **Definition 7.3**]). *For a weighted Dynkin diagram* $\Psi \in \mathrm{Map}(\Pi, \mathbb{R})$ *on* $D_{\mathfrak{g}}$ *(see Section 2.1 for the notation), we say that* Ψ *matches* $S_{(\mathfrak{g},\mathfrak{k})}$ *if all the weights on black nodes in* Π *are zero and any pair of nodes joined by an arrow have the same weights.*

Example 2.3. In the setting of Example 2.2, the weighted Dynkin diagram

$$\Phi = \begin{array}{c} a_5 \\ \overset{a_1\ a_2\ a_3}{\circ\!-\!\!\circ\!-\!\!\circ}\underset{a_4}{\diagdown} \\ \circ \end{array} \quad (a_1, \ldots, a_5 \in \mathbb{R})$$

on $D_{\mathfrak{o}(10)}$ matches the Satake diagram

$$S_{(\mathfrak{o}(10), \mathfrak{u}(5))} = \bullet\!-\!\!\circ\!-\!\!\bullet\!\!\diagdown\!\!\begin{array}{c}\circ\\ \circ\end{array}$$

if and only if $a_1 = a_3 = 0$ and $a_4 = a_5$.

3. Main results

3.1. *G-conjugacy classes of geodesics in Riemannian symmetric spaces*

In this subsection, we give a parametrization of Geod(M) for Riemannian symmetric space of compact type by weighted Dynkin diagrams (see Theorem 3.1 below).

Let $\mathfrak{g}$ be a compact semisimple Lie algebra and consider the Dynkin diagram $D_{\mathfrak{g}}$ of $\mathfrak{g}$ labelled by $\Pi = \{\alpha_1, \ldots, \alpha_r\}$, where $r = \mathrm{rank}\,\mathfrak{g}$. (see Section 2.1 for the notation).

To state our main results, we set up our notation as follows: For each $k \in \mathbb{R}_{>0}$ and non-trivial and non-negative weighted Dynkin diagram $\Phi \in \mathrm{Map}^*(\Pi, \mathbb{R}_{\geq 0})$ on $D_{\mathfrak{g}}$, let us define $k\Phi \in \mathrm{Map}^*(\Pi, \mathbb{R}_{\geq 0})$ by $(k\Phi)(\alpha) = k(\Phi(\alpha))$ for each $\alpha \in \Pi$. We denote by

$$[\Phi]_{\mathbb{R}_{>0}} := \{t\Phi \mid t \in \mathbb{R}_{>0}\} \subset \mathrm{Map}^*(\Pi, \mathbb{R}_{\geq 0})$$

for each $\Phi \in \mathrm{Map}^*(\Pi, \mathbb{R}_{\geq 0})$ and put

$$\mathcal{P}_{\mathfrak{g}} := \left\{ \begin{array}{c} \text{Non-trivial and non-negative} \\ \text{weighted Dynkin diagrams on } D_{\mathfrak{g}} \end{array} \right\} \Big/ \mathbb{R}_{>0}$$

$$= \mathrm{Map}^*(\Pi_{\mathfrak{g}}, \mathbb{R}_{\geq 0})/\mathbb{R}_{>0}$$

$$= \{[\Phi]_{\mathbb{R}_{>0}} \mid \Phi \in \mathrm{Map}^*(\Pi, \mathbb{R}_{\geq 0})\}.$$

Let us fix a Riemannian symmetric space M of compact type with the Lie algebra of $\text{Isom}(M)$ is isomorphic to $\mathfrak{g}$, and we denote by G the identity component of $\text{Isom}(M)$ as before. For each point p in M, we denote by K_p and $\mathfrak{k}_p$ the isotropy subgroup of G at p and its Lie algebra, respectively. Then $(\mathfrak{g}, \mathfrak{k}_p)$ is a symmetric pair. We denote by $S_{(\mathfrak{g}, \mathfrak{k}_p)}$ the Satake diagram of $(\mathfrak{g}, \mathfrak{k}_p)$ labelled by Π. It should be remarked that if we take other point p' of M, then the isotropy subgroup $K_{p'}$ of G at p' is G-conjugate to K. Hence, by the arguments in Section 2.2, the Satake diagram $S_{(\mathfrak{g}, \mathfrak{k}_p)}$ and $S_{(\mathfrak{g}, \mathfrak{k}_{p'})}$ (labelled by Π) are the same. We define *the Satake diagram S_M associated to M* as $S_{(\mathfrak{g}, \mathfrak{k}_p)}$ for a point $p \in M$.

Let us fix a maximal geodesic $\ell : \mathbb{R} \to M$ and we put $p := \ell(0)$. Recall that the tangent space $T_p M$ is naturally identified with $\mathfrak{s}_p$, where $\mathfrak{g} = \mathfrak{k}_p + \mathfrak{s}_p$ is the decomposition of $\mathfrak{g}$ associated to the symmetric pair $(\mathfrak{g}, \mathfrak{k}_p)$ defined above. Therefore, $A_\ell := \dot{\ell}(0)$ can be considered as an element of $\mathfrak{s}_p$. Then there exists a $(\mathfrak{a}_0, \mathfrak{t}_0, \Pi_0) \in \Theta_{(\mathfrak{g}, \mathfrak{k}_p)}$ such that $\sqrt{-1} A_\ell \in (\sqrt{-1}\mathfrak{a}_0)^+$ where $(\sqrt{-1}\mathfrak{a}_0)^+$ denotes the closed Weyl chamber with respect to Π_0, that is,

$$(\sqrt{-1}\mathfrak{a}_0)^+ := \{X \in \sqrt{-1}\mathfrak{a}_0 \mid \alpha(X) \geq 0 \text{ for any } \alpha \in \Pi_0\}.$$

For the simplicity, we put $\alpha_i^0 := \iota_{\Pi_0}(\alpha_i) \in \Pi_0$ for each $\alpha_i \in \Pi$. We define the weighted Dynkin diagram Φ_ℓ associated to ℓ by

$$\Phi_\ell : \Pi \to \mathbb{R}_{\geq 0}, \quad \alpha_i \mapsto \alpha_i^0(\sqrt{-1}A_\ell).$$

One can prove that Φ_ℓ does not depend on the choice of $(\mathfrak{a}_0, \mathfrak{t}_0, \Pi_0) \in \Theta_{(\mathfrak{g}, \mathfrak{k}_p)}$.

In this situation, the set of G-conjugacy classes of maximal geodesics $\text{Geod}(M)$ in M (see Section 1 for the notation) can be considered as a subset of $\mathcal{P}_\mathfrak{g}$ as follows (the detailed proof will be reported elsewhere).

Theorem 3.1. *The map*

$$\mathcal{P} : \text{Geod}(M) \to \mathcal{P}_\mathfrak{g}, [\ell]_G \mapsto [\Phi_\ell]_{\mathbb{R}_{>0}}$$

is injective, and the image $\mathcal{P}(\text{Geod}(M))$ of $\mathcal{P}$ can be determined as follows:

$$\mathcal{P}(\text{Geod}(M)) = \{[\Phi]_{\mathbb{R}_{>0}} \in \mathcal{P}_\mathfrak{g} \mid \Phi \text{ matches } S_M\}.$$

Example 3.1. Let us consider $\mathbb{R}^{10}$ as a real vector space of 10-dimension with the standard inner-product. For an integer k with $1 \leq k \leq 5$, we denote by

$$M = \text{Gr}_k^{\mathbb{R}}(\mathbb{R}^{10}) = \{k\text{-dimensional real linear subspaces of } \mathbb{R}^{10}\}$$

the Grassmannian manifold of rank k in $\mathbb{R}^{10}$. Then $M = \mathrm{Gr}_k(\mathbb{R}^{10})$ can be considered as a Riemannian symmetric space of compact type. In this situation, the identity component G of $\mathrm{Isom}(M)$ can be considered as $G = SO(10)/\{\pm I_{10}\}$ acting on $\mathrm{Gr}_k^{\mathbb{R}}(\mathbb{R}^{10})$, where I_{10} denotes the identity matrix of size 10.

Recall that the Dynkin diagram $D_{\mathfrak{g}}$ for $\mathfrak{g} = \mathfrak{o}(10)$ is

and thus $\mathcal{P}_{\mathfrak{g}} = \mathcal{P}_{\mathfrak{o}(10)}$ is

$$
\mathcal{P}_{\mathfrak{o}(10)} = \left\{\;
\begin{array}{c}
a_5 \\
a_1\; a_2\; a_3 \diagup \\
\circ\!-\!\circ\!-\!\circ \diagdown a_4 \\
\end{array}
\;\middle|\;
\begin{array}{l}
a_i \in \mathbb{R}_{\geq 0}(i = 1,\ldots,5), \\
(a_1,\ldots,a_5) \neq (0,\ldots,0)
\end{array}
\;\right\} \Big/ \mathbb{R}_{>0}
$$

By Theorem 3.1, let us classify G-conjugacy classes of $M = \mathrm{Gr}_k(\mathbb{R}^{10})$ as follows. The symmetric pair of Lie algebras for $p \in \mathrm{Gr}_k(\mathbb{R}^{10})$ is $(\mathfrak{g}, \mathfrak{k}_p) \simeq (\mathfrak{o}(10), \mathfrak{o}(k) \oplus \mathfrak{o}(10 - k))$. Then the Satake diagram S_M associated to $M = \mathrm{Gr}_k(\mathbb{R}^{10})$ and the parameter space $\mathcal{P}(\mathrm{Geod}(\mathrm{Gr}_k(\mathbb{R}^{10})))$ of G-conjugacy classes of geodesics in M (which is a subset of $\mathcal{P}_{\mathfrak{o}(10)}$) is in Table 1 below.

k	$S_{\mathrm{Gr}_k(\mathbb{R}^{10})}$	$\mathcal{P}(\mathrm{Geod}(\mathrm{Gr}_k(\mathbb{R}^{10}))) \subset \mathcal{P}_{\mathfrak{o}(10)}$
1		$\{a_2 = a_3 = a_4 = a_5 = 0\}/\mathbb{R}_{>0} \subset \mathcal{P}_{\mathfrak{o}(10)}$
2		$\{a_3 = a_4 = a_5 = 0\}/\mathbb{R}_{>0} \subset \mathcal{P}_{\mathfrak{o}(10)}$
3		$\{a_4 = a_5 = 0\}/\mathbb{R}_{>0} \subset \mathcal{P}_{\mathfrak{o}(10)}$
4		$\{a_4 = a_5\}/\mathbb{R}_{>0} \subset \mathcal{P}_{\mathfrak{o}(10)}$
5		$\mathcal{P}(\mathrm{Geod}(\mathrm{Gr}_5(\mathbb{R}^{10}))) = \mathcal{P}_{\mathfrak{o}(10)}$

Table 1: List of $\mathcal{P}(\mathrm{Geod}(\mathrm{Gr}_k(\mathbb{R}^{10})))$

3.2. Geodesics included in reflective submanifolds

Let us consider the setting in the previous subsection. We fix a reflective submanifold L of M. By the definition, L is a conected component of the set of all fixed points of an involutive isometry on M. The purpose of this paper is to determine $\mathcal{P}(\mathrm{Geod}(M;L))$ as a subset of $\mathcal{P}(\mathrm{Geod}(M))$ for each reflective submanifold L of M (see Section 1 for the notation of $\mathrm{Geod}(M;L)$ and see Theorem 3.1 for the notation of the map $\mathcal{P}: \mathrm{Geod}(M) \to \mathcal{P}_{\mathfrak{g}}$).

Let us put $G_L := \{g \in G \mid gL = L\}$ and denote by $\mathfrak{g}_L$ the Lie algebra of G_L. Then $(\mathfrak{g}, \mathfrak{g}_L)$ is a symmetric pair (see [9] for the details). We define the Satake diagram $S_{(M,L)}$ (labeled by Π) associated to the reflective submanifold L in M as the Satake diagram $S_{(\mathfrak{g}, \mathfrak{g}_L)}$ labeled by Π.

Then the subset $\mathcal{P}(\mathrm{Geod}(M;L))$ of $\mathcal{P}(\mathrm{Geod}(M))$ can be determined as follows (the detailed proof will be reported elsewhere):

Theorem 3.2.

$$\mathcal{P}(\mathrm{Geod}(M;L)) = \{\Phi \in \mathcal{P}(\mathrm{Geod}(M)) \mid \Phi \text{ matches } S_{(M,L)}\}$$
$$= \{\Phi \in \mathcal{P}_{\mathfrak{g}} \mid \Phi \text{ matches both } S_M \text{ and } S_{(M,L)}\}.$$

Example 3.2. Let us take $M = \mathrm{Gr}_k(\mathbb{R}^{10})$ as in Example 3.1. We fix a complex structure J on $\mathbb{R}^{10}$, that is, J is a linear endomorphism on $\mathbb{R}^{10}$ with $J \circ J = -\mathrm{id}_{\mathbb{R}^{10}}$. Then

$$L_J := \{V \in \mathrm{Gr}_k(\mathbb{R}^{10}) \mid J(V) = V\}$$

is a reflective submanifold of $\mathrm{Gr}_k(\mathbb{R}^{10})$. Note that $L_J = \emptyset$ if k is odd and

$$L_J \simeq \mathrm{Gr}_{k/2}(\mathbb{C}^5) := \{k/2\text{-dimensional complex linear subspaces of } \mathbb{C}^5\}$$

as a Riemannian symmetric space if k is even.

Let us take $k = 2, 4$ (even). Then the correspondence symmetric pair of $L_J \subset \mathrm{Gr}_k(\mathbb{R}^{10})$ is $(\mathfrak{g}, \mathfrak{g}_{L_J}) = (\mathfrak{o}(10), \mathfrak{u}(5))$. The Satake diagram $S_{(\mathrm{Gr}_k(\mathbb{R}^{10}), L_J)}$ associated to the reflective submanifold $\mathrm{Gr}_{k/2}(\mathbb{C}^5) \simeq L_J$ in $\mathrm{Gr}_k(\mathbb{R}^{10})$ is the following:

and by Theorem 3.2, the parameter space $\mathcal{P}(\mathrm{Geod}(\mathrm{Gr}_k(\mathbb{R}^{10}); L_J)) \subset \mathcal{P}(\mathrm{Geod}(\mathrm{Gr}_k(\mathbb{R}^{10})))$ can be determined as in Table 2 below.

k	$\mathcal{P}(\mathrm{Geod}(\mathrm{Gr}_k(\mathbb{R}^{10}); L_J)) \subset \mathcal{P}(\mathrm{Geod}(\mathrm{Gr}_k(\mathbb{R}^{10}))) \subset \mathcal{P}_{\mathfrak{o}(10)}$
2	$\{a_1 = a_3 = a_4 = a_5 = 0\}/\mathbb{R}_{>0} \subset \{a_3 = a_4 = a_5 = 0\}/\mathbb{R}_{>0} \subset \mathcal{P}_{\mathfrak{o}(10)}$
4	$\{a_1 = a_3 = 0,\ a_4 = a_5\}/\mathbb{R}_{>0} \subset \{a_4 = a_5\}/\mathbb{R}_{>0} \subset \mathcal{P}_{\mathfrak{o}(10)}$

Table 2: List of $\mathcal{P}(\mathrm{Geod}(\mathrm{Gr}_k(\mathbb{R}^{10}); L_J))$ as subsets of $\mathcal{P}(\mathrm{Geod}(\mathrm{Gr}_k(\mathbb{R}^{10})))$

Example 3.3. Let us take $M = \mathrm{Gr}_k(\mathbb{R}^{10})$ as in Example 3.1. We fix an integer t with $5 \leq t \leq 9$ and a subspace W of $\mathbb{R}^{10}$ with $\dim W = t$. We also fix an integer l with $0 \leq l \leq t$ and $k - l \leq 10 - t$. Then

$$L'_{W,l} := \left\{ V_1 \oplus V_2 \in \mathrm{Gr}_k(\mathbb{R}^{10}) \,\middle|\, \begin{array}{l} V_1 \subset W,\ V_2 \subset W^{\perp} \\ \text{with } \dim V_1 = l, \dim V_2 = k - l \end{array} \right\}$$

is a reflective submanifold of $\mathrm{Gr}_k(\mathbb{R}^{10})$. Note that

$$L'_{W,l} \simeq \mathrm{Gr}_l(\mathbb{R}^t) \times \mathrm{Gr}_{k-l}(\mathbb{R}^{10-t})$$

as a Riemannian symmetric space.

The correspondence symmetric pair of the reflective submanifold $L'_{W,l}$ of $\mathrm{Gr}_k(\mathbb{R}^{10})$ is $(\mathfrak{g}, \mathfrak{g}_{L'_{W,l}}) = (\mathfrak{o}(10), \mathfrak{o}(10 - t) \oplus \mathfrak{o}(t))$. The Satake diagram $S_{(\mathrm{Gr}_k(\mathbb{R}^{10}), L'_{W,l})}$ associated to the reflective submanifold $\mathrm{Gr}_l(\mathbb{R}^t) \times \mathrm{Gr}_{k-l}(\mathbb{R}^{10-t}) \simeq L'_{W,l}$ of $\mathrm{Gr}_k(\mathbb{R}^{10})$ is in Table 3 below.

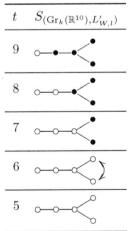

t	$S_{(\mathrm{Gr}_k(\mathbb{R}^{10}), L'_{W,l})}$

Table 3: List of Satake diagrams associated to the reflective submanifold $\mathrm{Gr}_l(\mathbb{R}^t) \times \mathrm{Gr}_{k-l}(\mathbb{R}^{10-t}) \simeq L'_{W,l}$ of $\mathrm{Gr}_k(\mathbb{R}^{10})$

Therefore, we find that the parameter space $\mathcal{P}(\mathrm{Geod}(\mathrm{Gr}_k(\mathbb{R}^{10}); L'_{W,l}))$ in $\mathcal{P}(\mathrm{Geod}(\mathrm{Gr}_k(\mathbb{R}^{10})))$ does not depend on the choice of l. The Table 4 below is the list of $\mathcal{P}(\mathrm{Geod}(\mathrm{Gr}_k(\mathbb{R}^{10}); L'_{W,l}))$ (see Example 3.1 for $\mathcal{P}(\mathrm{Geod}(\mathrm{Gr}_k(\mathbb{R}^{10})))$).

(k, t)	$\mathcal{P}(\mathrm{Geod}(\mathrm{Gr}_k(\mathbb{R}^{10}); L'_{W,l}))$
$k \leq 10 - t$	$\mathcal{P}(\mathrm{Geod}(\mathrm{Gr}_k(\mathbb{R}^{10}); L'_{W,l})) = \mathcal{P}(\mathrm{Geod}(\mathrm{Gr}_k(\mathbb{R}^{10})))$
$k > 10 - t$ and $t = 9$	$\{a_2 = a_3 = a_4 = a_5 = 0\} / \mathbb{R}_{>0}$
$k > 10 - t$ and $t = 8$	$\{a_3 = a_4 = a_5 = 0\} / \mathbb{R}_{>0}$
$k > 10 - t$ and $t = 7$	$\{a_4 = a_5 = 0\} / \mathbb{R}_{>0}$
$k > 10 - t$ and $t = 6$	$\{a_4 = a_5\} / \mathbb{R}_{>0}$

Table 4: List of $\mathcal{P}(\mathrm{Geod}(\mathrm{Gr}_k(\mathbb{R}^{10}); L'_{W,l}))$ as subsets $\mathcal{P}(\mathrm{Geod}(\mathrm{Gr}_k(\mathbb{R}^{10})))$

Remark 3.1. For $M = \mathrm{Gr}_5(\mathbb{R}^{10})$, there exists other types of reflective submanifolds than Examples 3.2 and 3.3. We omit the detailed here (see Leung [5, 6] for the details).

4. Applications

Let M be a Riemannian symmetric space of compact type and we denote by G the identity component of $\mathrm{Isom}(M)$ as before. We consider the following problem:

Problem 4.1. *Classify all pairs (L_1, L_2) of reflective submanifolds of M such that the intersection L_1 and gL_2 is discrete in M for any $g \in G$.*

Remark 4.1. If M is a compact Hermitian symmetric space and (L_1, L_2) are pair of real forms of M. Then there exists $g \in G$ such that $L_1 \cap gL_2$ is not discrete in M. Ikawa–Tanaka–Tasaki [3] gives an algorithm to classify $g \in G$ such that $L_1 \cap gL_2$ is discrete in M in terms of symmetric triad introduced by Ikawa [2].

For a pair (L_1, L_2) of complete totally geodesic submanifolds of M, if $L_1 \cap gL_2$ is not discrete in M, then the intersection $L_1 \cap gL_2$ includes a maximal geodesic of M. Therefore, we have the following observation:

Observation 4.1. For a pair (L_1, L_2) of complete totally geodesic submanifolds of M, the following conditions on (L_1, L_2) are equivalent:

(i) $L_1 \cap gL_2$ is discrete in M for any $g \in G$.
(ii) $\mathrm{Geod}(M; L_1) \cap \mathrm{Geod}(M; L_2) = \emptyset$ in $\mathrm{Geod}(M)$.

By Theorems 3.1 and 3.2 and classification results of reflective submanifolds by Leung [5, 6], Problem 4.1 can be solved for each Riemannian symmetric space M of compact type. We give an example of it for the cases where $M = \mathrm{Gr}_4(\mathbb{R}^{10})$. Detailed computations will be reported elsewhere.

Example 4.1. Let us take $M = \mathrm{Gr}_4(\mathbb{R}^{10})$ as in Example 3.1. By Leung [5, 6], reflective submanifolds of M are of the form of L_J for some complex structure J on $\mathbb{R}^{10}$ or $L'_{W,l}$ for some real linear subspaces W of $\mathbb{R}^{10}$ and integers l with $0 \leq l \leq \dim W$ and $4 - l \leq 10 - t$ (see Examples 3.2 and 3.3 for the notation). By the arguments in Examples 3.2 and 3.3, one can determine the subset $\mathrm{Geod}(\mathrm{Gr}_4(\mathbb{R}^{10}); L)$ of $\mathrm{Geod}(\mathrm{Gr}_4(\mathbb{R}^{10}))$ for $L = L_J$ or $L'_{W,l}$ and see that the pair $(L_J, L'_{W,l})$ satisfies the equivalent conditions in Observation 4.1 if $\dim W = 9$ and $l = 3$ or 4. Recall that $L_J \simeq \mathrm{Gr}_2(\mathbb{C}^5)$ and $L_{W,l} \simeq \mathrm{Gr}_3(\mathbb{R}^9)$ (if $\dim W = 9$ and $l = 3$) and $L_{W,l} \simeq \mathrm{Gr}_4(\mathbb{R}^9)$ (if $\dim W = 9$ and $l = 4$) as Riemannian symmetric spaces. This gives a complete classification (without swapping of L_1 and L_2) of pairs of reflective submanifolds (L_1, L_2) of $\mathrm{Gr}_4(\mathbb{R}^{10})$ satisfying the equivalent conditions in Observation 4.1.

References

[1] S. Araki, On root systems and an infinitesimal classification of irreducible symmetric spaces, *J. Math. Osaka City Univ.* **13**, 1–34, (1962).

[2] O. Ikawa, The geometry of symmetric triad and orbit spaces of Hermann actions, *J. Math. Soc. Japan* **63**, 79–136, (2011).

[3] O. Ikawa, M. S. Tanaka and H. Tasaki, The fixed point set of a holomorphic isometry, the intersection of two real forms in a Hermitian symmetric space of compact type and symmetric triads, *Internat. J. Math.* **26**, 1541005, 32 pages, (2015).

[4] D. S. P. Leung, The reflection principle for minimal submanifolds of Riemannian symmetric spaces, *J. Differential Geom.* **8**, 153–160, (1973).

[5] ———, On the classification of reflective submanifolds of Riemannian symmetric spaces, *Indiana Univ. Math. J.* **24**, 327–339, (1974/1975).

[6] ———, Errata: "On the classification of reflective submanifolds of Riemannian symmetric spaces" (Indiana Univ. Math. J. **24** (1974/75), 327–339), *Indiana Univ. Math. J.* **24**, 1199, (1975).

[7] T. Okuda, Classification of semisimple symmetric spaces with proper $SL(2, \mathbb{R})$-actions, *J. Differential Geom.* **94**, 301–342, (2013).

[8] I. Satake, On representations and compactifications of symmetric Riemannian spaces, *Ann. of Math.* (2) **71**, 77–110, (1960).
[9] H. Tasaki, Geometry of reflective submanifolds in Riemannian symmetric spaces, *J. Math. Soc. Japan* **58**, 275–297, (2006).

Received December 21, 2016
Revised February 14, 2017

A DIFFERENTIAL GEOMETRIC VIEWPOINT
OF MIXED HODGE STRUCTURES

Hisashi KASUYA

Department of Mathematics,
Graduate School of Science, Osaka University,
Osaka, Japan.
E-mail: kasuya@math.sci.osaka-u.ac.jp

We construct the canonical $\mathbb{R}$-mixed Hodge structure on Sullivan's 1-minimal mode of the de Rham complex of a compact Kähler manifold. Unlike the Hodge structure on the cohomology of a compact Kähler manifold, such $\mathbb{R}$-mixed Hodge structure depends on the choice of a Kähler metric and functorial for holomorphic isometric maps.

1. Introduction

An $\mathbb{R}$-*Hodge structure* of weight n on a $\mathbb{R}$-vector space V is a finite decreasing filtration F^* on $V_{\mathbb{C}} = V \otimes \mathbb{C}$ such that

$$F^p(V_{\mathbb{C}}) \oplus \overline{F^{n+1-p}(V_{\mathbb{C}})} = V_{\mathbb{C}}$$

for each p or equivalently, a bigrading

$$V_{\mathbb{C}} = \bigoplus_{p+q=n} V^{p,q}$$

on the complexification $V_{\mathbb{C}} = V \otimes \mathbb{C}$ such that

$$\overline{V^{p,q}} = V^{q,p}.$$

We can find $\mathbb{R}$-Hodge structures on compact Kähler manifolds. Let M be a compact complex manifold with a Kähler metric g and $A^*(M)$ the real de Rham complex with the natural bigrading

$$A^n(M) \otimes \mathbb{C} = \bigoplus_{p+q=n} A^{p,q}(M)$$

associated with the complex structure on M. By the decomposition $d = \partial + \bar{\partial}$ of the exterior derivative, $A^{*,*}(M)$ is a double complex. Define $\mathcal{H}_g^n(M) \subset A^n(M)$ by the space of the real harmonic n-forms on M associated with the metric g and $\mathcal{H}_g^{p,q}(M) \subset A^{p,q}(M)$ by the space of the (p,q)-forms which are harmonic. Then we have

$$\mathcal{H}_g^n(M) \otimes \mathbb{C} = \bigoplus_{p+q=n} \mathcal{H}_g^{p,q}(M)$$

and

$$\overline{\mathcal{H}_g^{p,q}(M)} = \mathcal{H}_g^{q,p}(M).$$

Hence, $\mathcal{H}_g^n(M)$ admits a $\mathbb{R}$-Hodge structure of weight n with the filtration

$$F^r(\mathcal{H}_g^n(M) \otimes \mathbb{C}) = \bigoplus_{p \geq r} \mathcal{H}_g^{p,q}(M).$$

By the natural isomorphism between $\mathcal{H}_g^n(M)$ and the de Rham cohomology $H^n(M, \mathbb{R})$, we obtain the $\mathbb{R}$-Hodge structure on $H^n(M, \mathbb{R})$ associated with the decomposition on $\mathcal{H}_g^n(M)$. We call the filtration of this $\mathbb{R}$-Hodge structure the *Hodge filtration* of $H^n(M, \mathbb{C})$. It is known that this $\mathbb{R}$-Hodge structure does not depend on the choice of a Kähler metric g. We can see this by the following way. Consider the spectral sequence of the double complex $A^{*,*}(M)$ (Frölicher spectral sequence). Then this spectral sequence degenerates at E_1-term and the filtration on $H^n(M, \mathbb{C})$ induced by this spectral sequence coincides with the Hodge filtration (see [13] for the detail). By this reason, Hodge structures are very important for complex analytic geometry and algebraic geometry but may not be so for differential geometry.

An $\mathbb{R}$-*mixed Hodge structure* is a generalization of a $\mathbb{R}$-Hodge structure. Let M be a compact Kähler manifold and D a normal crossing divisor. Then, for the complement $U = M - D$, Deligne show that the cohomology $H^n(U, \mathbb{R})$ admits a canonical $\mathbb{R}$-mixed Hodge structure and we can say that $\mathbb{R}$-mixed Hodge structures are invariants of smooth complex algebraic varieties ([3]).

The purpose of this survey is to study mixed Hodge theory for interests on differential geometry. We focus on $\mathbb{R}$-mixed Hodge structures on Sullivan's 1-minimal models of de Rham complexes of compact Kähler manifolds. In [9], Morgan constructed $\mathbb{R}$-mixed Hodge structures on Sullivan's 1-minimal models of de Rham complexes of complements $U = M - D$ as above. But, such structures are not unique determined and not functorial for holomorphic maps. In Section 5, we will suggest the canonical $\mathbb{R}$-mixed Hodge structures on Sullivan's 1-minimal models of de Rham complexes of Kähler manifolds. Such canonical $\mathbb{R}$-mixed Hodge structures depend on the choice of Kähler metrics and functorial for holomorphic isometric maps.

It is expected that this structure is applicable to differential geometry on Kähler manifolds.

2. Basics on mixed Hodge structures

Definition 2.1. An $\mathbb{R}$-*mixed Hodge structure* on an $\mathbb{R}$-vector space V is a pair (W_*, F^*) such that:

(1) W_* is an increasing filtration,
(2) F^* is a decreasing filtration on $V_{\mathbb{C}}$ such that the filtration on $Gr_n^W V_{\mathbb{C}} = W_n(V_{\mathbb{C}})/W_{n-1}(V_{\mathbb{C}})$ induced by F^* is an $\mathbb{R}$-Hodge structure of weight n.

We call W_* the weight filtration and F^* the Hodge filtration.

For two $\mathbb{R}$-vector spaces V_1 and V_2 with $\mathbb{R}$-mixed Hodge structures, a $\mathbb{R}$-linear map $\phi : V_1 \to V_2$ is a *morphism of $\mathbb{R}$-mixed Hodge structures* if $\phi(W_r(V_1)) \subset W_r(V_2)$ and $\phi(F^r(V_1 \otimes \mathbb{C})) \subset F^r(V_2 \otimes \mathbb{C})$ hold for any r.

Example 2.1. Let M be a compact Kähler manifold. Then for each n, the de Rham cohomology $H^n(M, \mathbb{R})$ admits an $\mathbb{R}$-Hodge structure of weight n. Consider the de Rham cohomology algebra

$$H^*(M, \mathbb{R}) = \bigoplus_{n=0}^{\dim_{\mathbb{R}} M} H^n(M, \mathbb{R}).$$

Take the filtrations (W_*, F^*) so that

$$W_r(H^*(M, \mathbb{R})) = \bigoplus_{n=0}^{r} H^n(M, \mathbb{R})$$

and

$$F^r(H^*(M, \mathbb{C})) = \bigoplus_{p \geq r} H^{p,q}(M).$$

Then, the pair (W_*, F^*) is a $\mathbb{R}$-mixed Hodge structure on $H^*(M, \mathbb{R})$.

Example 2.2 ([3] also [9, 11, 13]). *Let M be a compact Kähler manifold and D a normal crossing divisor. Consider the complement $U = M - D$. Then, for any $x \in M$, we have a neighborhood B which admits an isomorphism $B \cong \Delta^n$ so that $B \cap U \cong (\Delta^*)^l \times \Delta^{n-l}$ where Δ is the unit* *disc and* $\Delta^* = \Delta - \{0\}$. *We say that* $\alpha \in A^*(U)$ *is a* $\mathcal{C}^\infty$*-log form if for any such neighborhood* B *of each point, it is written as*

$$\alpha = \sum_{\{i_1,\ldots i_s\}\subset\{1,\ldots,l\}} \beta_{i_1,\ldots,i_s} \wedge \frac{dz_{i_1}}{z_{i_1}} \wedge \ldots \frac{dz_{i_1}}{z_{i_s}}$$

with $\beta_{i_1,\ldots,i_s} \in A^*(B)$ *for the coordinate* $B \cong \Delta^n$. *Define the sub-complex* $A^*(log(D)) \subset A^*(U) \otimes \mathbb{C}$ *of the* $\mathcal{C}^\infty$*-log forms. We define the filtration* W_* *of* $A^*(log(D))$ *so that* $\alpha \in W_r(A^*(log(D)))$ *is locally written as*

$$\alpha = \sum_{\{i_1,\ldots i_s\}\subset\{1,\ldots,l\}}^{s\leq r} \beta_{i_1,\ldots,i_s} \wedge \frac{dz_{i_1}}{z_{i_1}} \wedge \ldots \frac{dz_{i_1}}{z_{i_s}}.$$

We have the bigrading $A^n(log(D)) = \bigoplus_{p+q=n} A^{p,q}(log(D))$ *of* $A^*(log(D))$ *which is given by* $A^{*,*}(U)$. *But, we have*

$$A^{p,q}(log(D)) \neq \overline{A^{q,p}(log(D))}.$$

We define the filtration F^* *of* $A^*(log(D))$ *so that*

$$F^r(A^n(log(D))) = \bigoplus_{p\geq r} A^{p,q}(log(D)).$$

We can say that the inclusion $A^*(log(D)) \subset A^*(U) \otimes \mathbb{C}$ *induces a cohomology isomorphism and the filtration on* $H^n(U,\mathbb{C})$ *induced by* W_* *is "topological" and it is defined on* $H^n(U,\mathbb{R})$. *By this, in [3], by studying the spectral sequences of the filtrations, Deligne showed that the two filtration* (W'_*, F^*) *gives an* $\mathbb{R}$*-mixed Hodge structure on* $H^n(U,\mathbb{R})$ *for each n where* W'_* *is the shifted filtration (see Theorem 4.1).*

Proposition 2.1 ([3], [2], [9]). *Let* (W_*, F^*) *be an* $\mathbb{R}$*-mixed Hodge structure on an* $\mathbb{R}$*-vector space* V. *Define* $V^{p,q} = R^{p,q} \cap L^{p,q}$ *where*

$$R^{p,q} = W_{p+q}(V_{\mathbb{C}}) \cap F^p(V_{\mathbb{C}})$$

and

$$L^{p,q} = W_{p+q}(V_{\mathbb{C}}) \cap \overline{F^q(V_{\mathbb{C}})} + \sum_{i\geq 2} W_{p+q-i}(V_{\mathbb{C}}) \cap \overline{F^{q-i+1}(V_{\mathbb{C}})}.$$

Then we have the bigrading $V_{\mathbb{C}} = \bigoplus V^{p,q}$ *such that*

$$\overline{V^{p,q}} = V^{q,p} \mod \bigoplus_{r+s<p+q} V^{r,s},$$

$$W_i(V_{\mathbb{C}}) = \bigoplus_{p+q\leq i} V^{p,q} \qquad and \qquad F^i(V_{\mathbb{C}}) = \bigoplus_{p\geq i} V^{p,q}.$$

The bigrading $V_{\mathbb{C}} = \bigoplus V^{p,q}$ is called the bigrading of a $\mathbb{R}$-mixed Hodge structure (W_*, F^*).

We say that an $\mathbb{R}$-mixed Hodge structure (W_*, F^*) is split if $\overline{V^{p,q}} = V^{q,p}$. The $\mathbb{R}$-mixed Hodge structure in Example 2.1 is split.

3. Differential graded algebras and Sullivan's 1-minimal models

Let $\mathbb{K}$ be a field of characteristic 0.

Definition 3.1. A $\mathbb{K}$-*differential graded algebra* (called $\mathbb{K}$-DGA) is a graded $\mathbb{K}$-algebra A^* with the following properties:
(1) A^* is graded commutative, i.e.

$$y \wedge x = (-1)^{p \cdot q} x \wedge y \quad x \in A^p \quad y \in A^q.$$

(2) There is a differential operator $d : A \to A$ of degree one such that $d \circ d = 0$ and

$$d(x \wedge y) = dx \wedge y + (-1)^p x \wedge dy \quad x \in A^p.$$

Let A and B be $\mathbb{K}$-DGAs. If a morphism of graded algebra $\varphi : A \to B$ satisfies $d \circ \varphi = \varphi \circ d$, we call φ a morphism of $\mathbb{K}$-DGAs. If a morphism of $\mathbb{K}$-DGAs induces a cohomology isomorphism, we call it a quasi-isomorphism.

Example 3.1. The de Rham complex $A^*(M)$ of a smooth manifold M is an $\mathbb{R}$-DGA. For a smooth map $f : M \to N$, we have the morphism $f^* : A^*(N) \to A^*(M)$ of $\mathbb{R}$-DGAs.

Example 3.2. Let $\mathfrak{g}$ be a Lie algebra over $\mathbb{K}$. For the dual $d : \mathfrak{g}^* \to \mathfrak{g}^* \wedge \mathfrak{g}^*$ of the Lie bracket $\mathfrak{g} \wedge \mathfrak{g} \to \mathfrak{g}$, we extend d as the linear operator on the exterior algebra $\bigwedge \mathfrak{g}^*$. Then we have $d \circ d = 0$. Regarding $\mathfrak{g}^*$ as a graded vector space of degree 1 elements, $\bigwedge \mathfrak{g}^*$ is a $\mathbb{K}$-DGA. A Lie algebra homomorphism $f : \mathfrak{g} \to \mathfrak{h}$ induces the morphism $f^* : \bigwedge \mathfrak{h}^* \to \bigwedge \mathfrak{g}^*$ of $\mathbb{K}$-DGAs.

Conversely, if the exterior algebra $\bigwedge V$ with $deg(V) = 1$ admits a $\mathbb{K}$-DGA structure, then the dual $V^* \wedge V^* \to V^*$ of the differential operator $d : V \to V \wedge V$ is a Lie bracket.

Definition 3.2. Define the $\mathbb{K}$-DGA (t, dt) as the tensor product of the ring of $\mathbb{K}$-polynomials on t with the exterior algebra of $\langle dt \rangle$ so that t is of degree 0, $d(t) = dt$ and $d(dt) = 0$.

Let A^* be a $\mathbb{K}$-DGA. We consider the DGA $A^* \otimes (t, dt)$. Each element of $A^n \otimes (t, dt)$ can be written as

$$\sum (a_i t^i + b_i t^i dt)$$

with $a_i \in A^n$ and $b_i \in A^{n-1}$. We can define the "integrations" $\int_0^1 : A^* \otimes (t, dt) \to A^*$ and $\int_0^t : A^* \otimes (t, dt) \to A^* \otimes (t, dt)$ so that

$$\int_0^1 \sum (a_i t^i + b_i t^i dt) = (-1)^{n-1} \sum \frac{b_i}{i+1}$$

and

$$\int_0^t \sum (a_i t^i + b_i t^i dt) = (-1)^{n-1} \sum \frac{b_i t^{i+1}}{i+1}.$$

Definition 3.3. Let A^* and B^* be $\mathbb{K}$-DGAs. Let ϕ_0 and ϕ_1 be morphisms from A^* to B^*. Then, a homotopy from ϕ_0 to ϕ_1 is a morphism $H : A^* \to B^* \otimes (t, dt)$ so that for any $x \in A^*$ we have

$$H(x)|_{t=0} = \phi_0(x) \quad \text{and} \quad H(x)|_{t=1} = \phi_1(x).$$

See [5, Chapter 11] for basic properties of homotopies and integrations.

Definition 3.4. A $\mathbb{K}$-DGA $\mathcal{M}^*$ is 1-*minimal* if $\mathcal{M}^* = \bigcup \mathcal{M}_i^*$ for a sequence of sub-DGAs

$$\mathbb{K} = \mathcal{M}_0^* \subset \mathcal{M}_1^* \subset \mathcal{M}_2^* \subset \cdots$$

such that $\mathcal{M}_{i+1}^* = \mathcal{M}_i^* \otimes \bigwedge \mathcal{V}_{i+1}$, $\mathcal{V}_{i+1}$ is a graded vector space of degree 1 and $d\mathcal{V}_{i+1} \subset \mathcal{M}_i^2$.

For each i, we have

$$\mathcal{M}_i^* = \bigwedge (\mathcal{V}_1 \oplus \cdots \oplus \mathcal{V}_i).$$

Thus, by Example 3.2, the dual space $\mathcal{V}_1^* \oplus \cdots \oplus \mathcal{V}_i^*$ is a Lie algebra. We can easily check that this Lie algebra is nilpotent.

Definition 3.5. For a $\mathbb{K}$-DGA A^*, a 1-*minimal model* of A^* is a 1-minimal DGA $\mathcal{M}^*$ admitting a morphism $\phi : \mathcal{M}^* \to A^*$ which induces isomorphisms on 0-th and first cohomology and an injection on the second cohomology.

A $\mathbb{K}$-DGA A^* is *cohomologically connected* if $H^0(A^*) = \mathbb{K}$.

Theorem 3.1 ([4, 5]). *For any cohomologically connected* $\mathbb{K}$-*DGA* A^*, *a* 1-*minimal model of* A^* *exists and it is unique up to isomorphism of* $\mathbb{K}$-*DGA. More precisely, for any two* 1-*minimal models* $\mathcal{M}^*$ *and* $\mathcal{N}^*$ *with maps* $\phi : \mathcal{M}^* \to A^*$ *and* $\psi : \mathcal{N}^* \to A^*$ *inducing isomorphisms on* 0-*th and first cohomologies and an injections on the second cohomologies, there exists an isomorphism* $\mathcal{I} : \mathcal{M}^* \to \mathcal{N}^*$ *and a homotopy* $\mathcal{H} : \mathcal{M}^* \to A^* \otimes (t, dt)$ *from* $\psi \circ \mathcal{I}$ *to* ϕ.

By the uniqueness, we can say that if there exists a quasi-isomorphism between two cohomologically connected $\mathbb{K}$-DGA A_1^* and A_2^*, then they have the same 1-minimal model. Without the homotopy argument, a map $\phi : \mathcal{M}^* \to A^*$ is not unique.

For the existence of a 1-minimal model, we explain the inductive construction. Let $B_1 = \mathrm{ker}d_{|A^1}$. Take a sub-vector space $C^1, D^1 \subset A^1$ so that $A^1 = B^1 \oplus C^1$ and $B^1 = dA^0 \oplus D^1$ (by zorn's lemma). We define the map $\delta = (d_{|C^1})^{-1} : dA^1 \to C^1$. Then we construct $\mathcal{M}_i^* = \bigwedge(\mathcal{V}_1 \oplus \cdots \oplus \mathcal{V}_i)$ and the map $\phi_i : \mathcal{M}_i^* \to A^*$ by the following way.

(1) $\mathcal{V}_1 = D^1$. Define $\phi : \bigwedge \mathcal{V}_1 \to A^*$ so that ϕ is the natural injection $\mathcal{V}_1 = D^1 \hookrightarrow A^1$ on $\mathcal{V}_1$.
(2) $i \geq 1$. Let $L_i = \mathrm{ker}d_{|\mathcal{M}^2}$ and K_i be the kernel of the map $L_i \to H^2(A^*)$ given by $\phi : \mathcal{M}_i^* \to A^*$. Take $J_i \subset \mathcal{M}_2^*$ so that $K_i = d\mathcal{M}_i^1 \oplus J_i$. Then, we define $\mathcal{V}_{i+1} = J_i$ as $deg(\mathcal{V}_{i+1}) = 1$ and the differential operator d on $\bigwedge(\mathcal{V}_1 \oplus \cdots \oplus \mathcal{V}_{i+1})$ which is the natural injection $\mathcal{V}_{i+1} = J_i \hookrightarrow \mathcal{M}_2^*$. We extend $\phi_{i+1} : \bigwedge(\mathcal{V}_1 \oplus \cdots \oplus \mathcal{V}_{i+1}) \to A^*$ so that $\phi_{n+1}(v_{i+1}) = \delta \circ \phi_i(dv_{i+1})$ for $v_{i+1} \in \mathcal{V}_{i+1}$.

For the de Rham complex $A^*(M)$ of a smooth manifold M with a Riemannian metric g, we can take $D_1 = \mathcal{H}_g^1(M)$ and $\delta = d^*G$ where d^* is the adjoint operator of the differential d and G is the Green operator.

The construction of an isomorphism $\mathcal{I}$ and a Homotopy H is a consequence of the following result.

Proposition 3.1 ([5, Proposition 11.2]). *Let* A^*, B^* *and* C^* *be* $\mathbb{K}$-*DGAs. For a graded* $\mathcal{V}$ *with* $deg(\mathcal{V}) = 1$, *we assume that* $A^* \otimes \bigwedge \mathcal{V}$ *is a* $\mathbb{K}$-*DGA such that* $A^* \otimes \langle 1 \rangle$ *is a sub-DGA which is identified with* A^* *and* $d\mathcal{V} \subset A^2$. $\mathcal{I} : A^* \to B^*$, $\phi : A^* \otimes \bigwedge \mathcal{V} \to C^*$ *and* $\psi : B^* \to C^*$ *be morphisms such that there exists a homotopy* $H : A^* \to C^* \otimes (t, dt)$ *form* $\psi \circ \mathcal{I}$ *to* $\phi_{|A^*}$. *Then, if there exists a linear maps* $a : \mathcal{V} \to B^1$ *and* $b : \mathcal{V} \to C^0$ *so that*

$da(v) = \mathcal{I}(dv)$ *and*

$$db(v) = \phi(v) - \psi(a(v)) - \int_0^1 H(dv),$$

then defining the morphisms $\tilde{\mathcal{I}} : A^* \otimes \bigwedge \mathcal{V} \to B^*$ *and* $\tilde{H} : A^* \otimes \bigwedge \mathcal{V} \to$
$C^* \otimes (t, dt)$ *so that* $\tilde{\mathcal{I}}(v) = a(v)$ *and*

$$\tilde{H}(v) = \psi(\mathcal{I}(v)) + \int_0^t H(dv) + d(b(v)t),$$

the morphism $\tilde{H}$ *is a homotopy form* $\psi \circ \tilde{\mathcal{I}}$ *to* ϕ.

By this proposition, we can inductively construct $\mathcal{I}$ and H as in Theorem
3.1 (see [9, Section 5]).

We explain the important application of 1-minimal models to algebraic
topology established by Sullivan (see [12], [4], [5] for details). Let M be a
manifold with the finitely generated fundamental group $\pi_1(M, x)$. Consider
the tower of nilpotent groups

$$\cdots \to \Gamma_3 \to \Gamma_2 \to \Gamma_1 = \{e\}.$$

where Γ_i is the quotient of $\pi_1(M, x)$ by the i-the term of lower central
series. Then it is possible to "tensor" these nilpotent groups with a field $\mathbb{K}$
and we obtain the tower of real nilpotent Lie algebras

$$\cdots \to \mathfrak{n}_3 \to \mathfrak{n}_2 \to \{0\}.$$

The inverse limit of this tower is called the $\mathbb{K}$-Malcev Lie algebra of
$\pi_1(M, x)$. Let $\mathcal{M}^*$ be the 1-minimal model of $A^*(M)$. Then by the above
argument, $\mathcal{M}^*$ is corresponds to a pro-nilpotent Lie algebra (i.e. inverse
limit of nilpotent Lie algebras) $\mathfrak{m}$. Sullivan showed that $\mathfrak{m}$ is isomorphic to
the $\mathbb{R}$-Malcev Lie algebra of $\pi_1(M, x)$.

Remark 3.1. We also define the n-minimal model of a cohomologically
connected $\mathbb{K}$-DGA A^* for any $1 \le n \le +\infty$. We can also show the existence
and uniqueness of each n-minimal model The $+\infty$-minimal model is called
minimal model. For a simply connected manifold M, Sullivan showed that
the minimal model of the de Rham complex $A^*(M)$ corresponds to the real
homotopy $\pi_*(M, x) \otimes \mathbb{R}$ (see [4], [5] for details).

4. Morgan's mixed Hodge diagrams

Definition 4.1 ([9, Definition 3.5]). *An* $\mathbb{R}$-*mixed-Hodge diagram* *is a
pair of filtered* $\mathbb{R}$-*DGA* (A^*, W_*) *and bifiltered* $\mathbb{C}$-*DGA* (E^*, W_*, F^*) *and
filtered DGA map* $\phi : (A_\mathbb{C}^*, W_*) \to (E^*, W_*)$ *such that:*

(1) The map ϕ induces an isomorphism ϕ^* : $_W E_1^{*,*}(A_{\mathbb{C}}^*) \to {}_W E_1^{*,*}(E^*)$, where $_W E_*^{*,*}(\cdot)$ is the spectral sequence for the decreasing filtration $W^* = W_{-*}$.

(2) The differential d_0 on $_W E_0^{*,*}(E^*)$ is strictly compatible with the filtration induced by F.

(3) The filtration on $_W E_1^{p,q}(E^*)$ induced by F is an $\mathbb{R}$-Hodge structure of weight q on $\phi^*({}_W E_1^{*,*}(A^*))$.

Theorem 4.1 ([9, Theorem 4.3]). Let $\{(A^*, W_*), (E^*, W_*, F^*), \phi\}$ be an $\mathbb{R}$-mixed-Hodge diagram. Define the filtration W_*' on $H^r(A^*)$ (resp. $H^r(E^*)$) as $W_i'H^r(A^*) = W_{i-r}(H^r(A^*))$ (resp. $W_i'H^r(E^*) = W_{i-r}(H^r(E^*))$). Then the filtrations W_*' and F^* on $H^r(E^*)$ give an $\mathbb{R}$-mixed-Hodge on $H^r(A^*)$ via the isomorphism ϕ^* : $H^r(A_{\mathbb{C}}^*) \to H^r(E^*)$.

Theorem 4.2 ([9]). Let $\{(A^*, W_*), (E^*, W_*, F^*), \phi\}$ be an $\mathbb{R}$-mixed-Hodge diagram and $\mathcal{M}^*$ the 1-minimal model of A^* with a map $\phi : \mathcal{M}^* \to A^*$ inducing isomorphisms on 0-th and first cohomology and an injection on the second cohomology. Then, there exists an $\mathbb{R}$-mixed Hodge structure (W_*, F^*) on $\mathcal{M}^*$ such that the product, differential operator and the induced map $\phi : H^*(\mathcal{M}^*) \to H^*(A^*)$ are morphisms of $\mathbb{R}$-mixed Hodge structures where the $\mathbb{R}$-mixed Hodge structure on $H^r(A^*)$ is the one as in Theorem 4.1.

Remark 4.1. We give an overview of Morgan's construction. We construct:

(1) a bigrading $\mathcal{N}^* = \bigoplus \mathcal{N}^*(p, q)$ of the 1-minimal model $\psi : \mathcal{N}^* \to E^*$ of the $\mathbb{C}$-DGA E^* ([9, Section 6]),

(2) a filtration W_* on the 1-minimal model $\phi : \mathcal{M}^* \to A^*$ of the $\mathbb{R}$-DGA A^* ([9, Section 7]),

(3) a filtration preserving isomorphism $\mathcal{I} : (\mathcal{M}^* \otimes \mathbb{C}, W_*) \to (\mathcal{N}^*, W_*)$ with a homotopy from $\psi \circ \mathcal{I}$ to ϕ where

$$W_r(\mathcal{N}^*) = \bigoplus_{p+q \leq r} \mathcal{N}^*(p, q)$$

([9, Section 7]).

Then, for the filtration $\mathcal{F}^*$ on $\mathcal{M}^*$ so that

$$F^r(\mathcal{M}^*) = \mathcal{I}^{-1}\left(\bigoplus_{p \geq r} \mathcal{I}^{-1}\mathcal{M}^*(p, q)\right),$$

(W_*, F^*) is an $\mathbb{R}$-mixed Hodge structure on $\mathcal{M}^*$ ([9, Section 8]). Morgan showed that the bigrading $\mathcal{N}^* = \bigoplus \mathcal{N}^*(p,q)$ and the filtration W_* on $\mathcal{M}^*$ are uniquely determined by an $\mathbb{R}$-mixed-Hodge diagram $\{(A^*, W_*), (E^*, W_*, F^*), \phi\}$. But a filtration preserving isomorphism $\mathcal{I}$: $(\mathcal{M}^* \otimes \mathbb{C}, W_*) \to (\mathcal{N}^*, W_*)$ is not unique and an $\mathbb{R}$-mixed Hodge structure (W_*, F^*) on $\mathcal{M}^*$ depend on the choice of $\mathcal{I}$. Thus, an $\mathbb{R}$-mixed Hodge structure (W_*, F^*) on $\mathcal{M}^*$ is not unique.

Example 4.1. Let M be a compact Kähler manifold. We consider the de Rham cohomology algebra $H^*(M, \mathbb{R})$ (resp. $H^*(M, \mathbb{C})$ as a $\mathbb{R}$-DGA (resp. $\mathbb{C}$-DGA) with the trivial differential operator $d = 0$. Define the filtration W_* on $H^*(M, \mathbb{R})$ so that $W_r(H^*(M, \mathbb{R})) = 0$ for $r < 0$ and $W_r(H^*(M, \mathbb{R})) = H^*(M, \mathbb{R})$ for $r \geq 0$. The filtration F^* on $H^*(M, \mathbb{C})$ is the same as Example 2.1. Then, we can easily check that

$$\{(H^*(M, \mathbb{R}), W_*), (H^*(M, \mathbb{C}), W_*, F^*), \mathrm{id}\}$$

is an $\mathbb{R}$-mixed-Hodge diagram.

Let $\mathcal{M}^*$ be the 1-minimal model of $H^*(M, \mathbb{R})$. We can construct an $\mathbb{R}$-mixed Hodge structure on $\mathcal{M}^*$ explicitly. $\mathcal{M}^* = \bigwedge(\mathcal{V}_1 \oplus \mathcal{V}_2 \cdots \oplus \mathcal{V}_n \oplus \dots)$ is given by the following way see the inductive construction in Section 3:

(1) $\mathcal{V}_1 = H^1(M, \mathbb{R})$.
(2) $\mathcal{V}_2$ is the kernel of the cup product $H^1(M, \mathbb{R}) \wedge H^1(M, \mathbb{R}) \to H^2(M, \mathbb{R})$ the differential operator d is the natural injection $\mathcal{V}_2 \hookrightarrow \mathcal{V}_1 \wedge \mathcal{V}_1$.
(3) $n \geq 2$. $\mathcal{V}_{n+1} = \ker d |_{\sum_{i+j=n+1} \mathcal{V}_i \wedge \mathcal{V}_j}$.

Obviously, $\mathcal{V}_1$ admits a $\mathbb{R}$-Hodge structure of weight 1. It is known that the cup product $H^1(M, \mathbb{R}) \wedge H^1(M, \mathbb{R}) \to H^2(M, \mathbb{R})$ is a morphism of $\mathbb{R}$ Hodge structures and hence $\mathcal{V}_2$ admits a $\mathbb{R}$-Hodge structure of weight 2. Inductively, $\mathcal{V}_n$ admits a $\mathbb{R}$-Hodge structure of weight n for each n. Thus, we obtain the split $\mathbb{R}$-mixed Hodge structure on $\mathcal{M}^*$.

Example 4.2. Let M be a compact Kähler manifold. We consider the de Rham complex $A^*(M)$. Define the filtration W_* on $A^*(M)$ so that $W_r(A^*(M)) = 0$ for $r < 0$ and $W_r(A^*(M)) = A^*(M)$ for $r \geq 0$. We define the filtration F^* of $A^*(M) \otimes \mathbb{C}$ so that

$$F^r(A^n(M)) = \bigoplus_{p \geq r} A^{p,q}(M).$$

Then, we have $E_0^{0,q} = A^q(M)$ and $E_0^{p,q} = 0$ for any $p \neq 0$. As we explained in Introduction F^* induces a Hodge structure of weight q on $H^q(M, \mathbb{C})$.

Thus, we can say that

$$\{(A^*(M), W_*), (A^*(M) \otimes \mathbb{C}, W_*, F^*), \mathrm{id}\}$$

is an $\mathbb{R}$-mixed-Hodge diagram.

Example 4.3. Let M be a compact Kähler manifold and D a normal crossing divisor. We consider $(A^*(log(D)), W_*, F^*)$ as in Example 2.2. Then, $A^*(log(D))$ is closed under the wedge product and (W_*, F^*) are multiplicative and so $(A^*(log(D)), W_*, F^*)$ is a bifiltered $\mathbb{C}$-DGA. We notice that $A^*(log(D))$ does not admits a real structure.

In [9], Morgan constructed a filtered $\mathbb{R}$-DGA (A^*, W_*) and a filtration preserving quasi-isomorphism $\phi : (A^*, W_*) \to (A^*(log(D)), W_*)$ such that

$$\{(A^*, W_*), (A^*(log(D)), W_*, F^*), \phi\}$$

is an $\mathbb{R}$-mixed-Hodge diagram which induces Deligne's $\mathbb{R}$-mixed Hodge structure by using simplicial de Rham theory. Moreover, such $\mathbb{R}$-DGA A^* admits a quasi-isomorphism $A^* \to A^*(U)$. Thus, we can say that the 1-minimal model of the de Rham complex $A^*(U)$ admits an $\mathbb{R}$-mixed Hodge structure and by Sullivan's theorem the $\mathbb{R}$-Malcev Lie algebra of the fundamental group of U admits an $\mathbb{R}$-mixed Hodge structure. By this $\mathbb{R}$-mixed Hodge structure, we can say that certain finitely generated groups can not be the fundamental group of any smooth algebraic variety. (See [9, Corollary 10.3, 10.4], also [1, Chapter 3.6].)

5. $\mathbb{R}$-mixed Hodge structures and Kähler metrics

By Example 4.2, the 1-minimal model of the de Rham complex of a compact Kähler manifold admits an $\mathbb{R}$-mixed Hodge structure. But such $\mathbb{R}$-mixed Hodge structure is not unique. In this section we will construct the canonical one.

Let M be a compact complex manifold with a Kähler metric g. On the de Rham complex $A^*(M)$ with the bigrading $A^n(M) \otimes \mathbb{C} = \bigoplus_{p+q=n} A^{p,q}(M)$, we consider the differential operators d, ∂, $\bar{\partial}$ and $d^c = -\sqrt{-1}(\partial - \bar{\partial})$ their adjoints d^*, ∂^*, $\bar{\partial}^*$ and d^{c*} associated with g, their Laplacian operators Δ, Δ^c, Δ' and Δ'' and their Green operators G, G^c, G' and G''. Then we have

$$\Delta = \Delta^c = 2\Delta' = 2\Delta''.$$

We define

$$F^c = d^* G d^{c*} G^c \qquad \text{and} \qquad F' = -2\sqrt{-1}F^c.$$

Then we have the following relations (see [4, Lemma 5.11 and its proof]).

Proposition 5.1.

$(dd^c$-**Lemma**$)$ *For* $\alpha \in \operatorname{im}d \cap \operatorname{ker}d^c$, *we have* $\alpha = dd^c F^c \alpha$ *and hence*

$$\operatorname{im}d \cap \operatorname{ker}d^c = \operatorname{ker}d \cap \operatorname{im}d^c = \operatorname{im}dd^c.$$

$(\partial\bar{\partial}$-**Lemma**$)$ *For* $\alpha \in \operatorname{im}\partial \cap \operatorname{ker}\bar{\partial}$, *we have* $\alpha = \partial\bar{\partial}F'\alpha$ *and hence*

$$\operatorname{im}\partial \cap \operatorname{ker}\bar{\partial} = \operatorname{ker}\partial \cap \operatorname{im}\bar{\partial} = \operatorname{im}\partial\bar{\partial}.$$

We consider the sub-DGA $\operatorname{ker}d^c \subset A^*(M)$ (resp. $\operatorname{ker}\partial \subset A^*(M) \otimes \mathbb{C}$). We regard the cohomologies $H^*_{d^c}(M)$ and $H^*_{\partial}(M)$ as DGAs with the differential operators d. By these two relations, we have the following (see [4, Section 6]).

Theorem 5.1.

- *The inclusions*

$$\operatorname{ker}d^c \subset A^*(M) \quad \text{and} \quad \operatorname{ker}\partial \subset A^*(M) \otimes \mathbb{C}$$

  are quasi-isomorphisms.
- *The quotients*

$$\operatorname{ker}d^c \to H^*_{d^c}(M) \quad \text{and} \quad \operatorname{ker}\partial \to H^*_{\partial}(M)$$

  are quasi-isomorphisms.
- $d = 0$ *on* $H^*_{d^c}(M)$ *and* $H^*_{\partial}(M)$ *and hence* $H^*_{d^c}(M) \cong H^*(M,\mathbb{R})$ *and* $H^*_{\partial}(M) \cong H^*(M,\mathbb{C})$.

By this theorem, we can say that two DGAs $A^*(M)$ and $H^*(M,\mathbb{R})$ have the same 1-minimal model. Thus, by the argument in Example 4.1, the 1-minimal model of $A^*(M)$ admits a sprit $\mathbb{R}$-mixed Hodge structure. But, this $\mathbb{R}$-mixed Hodge structure is not interesting, because it is determined by the $\mathbb{R}$-Hodge structure on $H^1(M,\mathbb{R})$ and the cup product $H^1(M,\mathbb{R}) \wedge H^1(M,\mathbb{R}) \to H^2(M,\mathbb{R})$.

We will construct a canonical $\mathbb{R}$-mixed Hodge structure on the 1-minimal model of $A^*(M)$ associated with a Kähler metric g. We first construct the canonical 1-minimal models $\phi : \mathcal{M}^* \to A^*(M)$ and $\psi : \mathcal{N}^* \to A^*(M) \otimes \mathbb{C}$.

Construction 5.1. We construct $\mathcal{M}^*_i = \bigwedge(\mathcal{V}_1 \oplus \cdots \oplus \mathcal{V}_i)$ and the map $\phi_i : \mathcal{M}^*_i \to A^*(M)$ by the following way.

(1) $\mathcal{V}_1 = \mathcal{H}^1(M) = \ker d \cap \ker d^c \cap A^1(M)$. Define $\phi_1 : \bigwedge \mathcal{V}_1 \to A^*(M)$ so that ϕ is the natural injection $\mathcal{V}_1 = \mathcal{H}^1(M) \hookrightarrow A^1(M)$ on $\mathcal{V}_1$.

(2) Define $\mathcal{V}_2$ as the kernel of the cup product

$$\bigwedge^2 \mathcal{V}_1 \to H^2(M, \mathbb{R}).$$

Define the DGA $\mathcal{M}_2^* = \bigwedge(\mathcal{V}_1 \oplus \mathcal{V}_2)$ with the differential d so that d is 0 on $\mathcal{V}_1$ and d on $\mathcal{V}_2$ is the natural inclusion $\mathcal{V}_2 \hookrightarrow \bigwedge^2 \mathcal{V}_1$. Extend the morphism $\phi_2 : \mathcal{M}_2^* \to A^*(M)$ so that $\phi_2(v) = d^c F^c(\phi_1(dv))$ for $v \in \mathcal{V}_2$.

(3) For $i \geq 2$, consider the DGA $\mathcal{M}_i^* = \bigwedge(\mathcal{V}_1 \oplus \mathcal{V}_2 \oplus \cdots \oplus \mathcal{V}_i)$ with the homomorphism $\phi_i : \mathcal{M}_i^* \to A^*(M)$ we have constructed. We have $\phi_i(v) \in \mathrm{im} d^c$ for $v \in \mathcal{V}_2 \oplus \cdots \oplus \mathcal{V}_i$.

Let

$$\mathcal{V}_{i+1} = \ker d|_{\sum_{j+k=i+1} \mathcal{V}_j \wedge \mathcal{V}_k}.$$

Define the extended DGA $\mathcal{M}_{i+1}^* = \mathcal{M}_i^* \otimes \bigwedge \mathcal{V}_{i+1}$ so that the differential d is defined on $\mathcal{V}_{i+1}$ as the natural inclusion $\mathcal{V}_{i+1} \hookrightarrow \sum_{j+k=i+1} \mathcal{V}_j \wedge \mathcal{V}_k$. Extend the morphism $\phi_{i+i} : \mathcal{M}_{i+1}^* \to A^*(M)$ is defined by $\phi_{i+1}(v) = d^c F^c(\phi_i(dv))$ for $v \in \mathcal{V}_{i+1}$.

We have the grading $\mathcal{M}^* = \bigoplus \mathcal{M}^*(n)$ which commutes with the multiplication and the differential operator so that $\mathcal{M}^1(i) = \mathcal{V}_i$.

Construction 5.2. We construct $\mathcal{N}_i^* = \bigwedge(\mathcal{W}_1 \oplus \cdots \oplus \mathcal{W}_i)$ and the map $\psi_i : \mathcal{N}_i^* \to A^*(M) \otimes \mathbb{C}$ by the following way.

(1) $\mathcal{W}_1 = \mathcal{H}^1(M) \otimes \mathbb{C} = \ker \partial \cap \ker \bar{\partial} \cap A^1(M) \otimes \mathbb{C}$. Define $\psi_1 : \bigwedge \mathcal{W}_1 \to A^*(M) \otimes \mathbb{C}$ so that ψ_1 is the natural injection $\mathcal{V}_1 = \mathcal{H}^1(M) \hookrightarrow A^1(M) \otimes \mathbb{C}$ on $\mathcal{W}_1$.

(2) Define $\mathcal{W}_2$ as the kernel of the cup product

$$\bigwedge^2 \mathcal{W}_1 \to H^2(M, \mathbb{C}).$$

Define the DGA $\mathcal{N}_2^* = \bigwedge(\mathcal{W}_1 \oplus \mathcal{W}_2)$ with the differential d so that d is 0 on $\mathcal{W}_1$ and d on $\mathcal{W}_2$ is the natural inclusion $\mathcal{W}_2 \hookrightarrow \bigwedge^2 \mathcal{W}_1$. Extend the morphism $\psi_2 : \mathcal{N}_2^* \to A^*(M) \otimes \mathbb{C}$ so that $\psi_2(w) = \partial F'(\psi_1(dw))$ for $w \in \mathcal{W}_2$.

(3) For $i \geq 2$, consider the DGA $\mathcal{N}_i^* = \bigwedge(\mathcal{W}_1 \oplus \mathcal{W}_2 \oplus \cdots \oplus \mathcal{W}_i)$ with the homomorphism $\psi_i : \mathcal{N}_i^* \to A^*(M) \otimes \mathbb{C}$ we have constructed. We have $\psi_i(w) \in \mathrm{im} \partial$ for $w \in \mathcal{W}_2 \oplus \cdots \oplus \mathcal{W}_i$.

Let
$$\mathcal{W}_{i+1} = \ker d |_{\sum_{j+k=i+1} \mathcal{W}_j \wedge \mathcal{W}_k}.$$
Define the extended DGA $\mathcal{N}_{i+1}^* = \mathcal{N}_i^* \otimes \bigwedge \mathcal{W}_{i+1}$ so that the differential d is defined on $\mathcal{W}_{i+1}$ as the natural inclusion $\mathcal{W}_{i+1} \hookrightarrow \sum_{j+k=i+1} \mathcal{W}_j \wedge \mathcal{W}_k$. Extend the morphism $\psi_{i+i} : \mathcal{N}_{i+1}^* \to A^*(M) \otimes \mathbb{C}$ is defined by $\psi_{i+1}(w) = \partial F'(\psi_i(dw))$ for $w \in \mathcal{W}_{i+1}$.

We have the grading $\mathcal{N}^* = \bigoplus \mathcal{N}^*(n)$ which commutes with the multiplication and the differential operator so that $\mathcal{N}^1(i) = \mathcal{W}_i$. For the decomposition $\mathcal{H}^1(M) \otimes \mathbb{C} = \mathcal{H}^{1,0}(M) \oplus \mathcal{H}^{0,1}(M)$, we define $\mathcal{W}^{1,0} = \mathcal{H}^{1,0}(M)$, $\mathcal{W}^{0,1} = \mathcal{H}^{0,1}(M)$ and
$$\mathcal{W}^{p,q} = \ker d |_{\sum_{s+u=p,t+v=Q} \mathcal{W}^{s,t} \wedge \mathcal{W}^{u,v}}$$
inductively. Then we have $\mathcal{W}_n = \bigoplus_{p+q=n} \mathcal{W}^{p,q}$. By this, we have the bigrading $\mathcal{N}^* = \bigoplus \mathcal{N}^*(p,q)$ which commutes with the multiplication and the differential operator so that $\mathcal{N}^1(p,q) = \mathcal{W}^{p,q}$.

These constructions give well-defined 1-minimal models $\phi : \mathcal{M}^* \to A^*(M)$ and $\psi : \mathcal{N}^* \to A^*(M)$ by dd^c-Lemma and $\partial\bar{\partial}$-Lemma.

We will construct the $\mathbb{R}$-mixed Hodge structure by using the above 1-minimal models. Define the filtrations
$$W_k(\mathcal{M}^*) = \bigoplus_{i \le k} \mathcal{M}^*(i) \quad \text{and} \quad W_k(\mathcal{N}^*) = \bigoplus_{i \le k} \mathcal{N}^*(i).$$

Construction 5.3. We construct the filtration preserving isomorphism $\mathcal{I} : (\mathcal{M}^* \otimes \mathbb{C}, W_*) \to (\mathcal{N}, W_*)$ and the homotopy $H : \mathcal{M}^* \otimes \mathbb{C} \to A^*(M) \otimes \mathbb{C} \otimes (t, dt)$ from $\psi \circ \mathcal{I}$ to ϕ by the following way.

(1) We have $\mathcal{M}_1^* \otimes \mathbb{C} = \mathcal{M}_1^*$ and $\phi_1 = \psi_1$. Define $\mathcal{I} = \mathrm{id}$ and $H = \phi_1$.
(2) $n \ge 1$. We assume that there exists $\mathcal{I}_n : \mathcal{M}_n^* \otimes \mathbb{C} \to \mathcal{N}_n^*$ and a homotopy $H : \mathcal{M}_n^* \otimes \mathbb{C} \to A^*(M) \otimes \mathbb{C} \otimes (t, dt)$ from ϕ_n to $\psi_n \circ \mathcal{I}$. For $v \in \mathcal{V}_{n+1}$, we can take a unique $a(v) \in \mathcal{N}_{n+1}^1$ so that $\mathcal{I}(dv) = da(v)$ and
$$\phi_{n+1}(v) - \psi_{n+1}(a(v)) - \int_0^1 H(dv)$$
is a exact form. Actually, since the map $d : \mathcal{W}_2 \oplus \ldots \mathcal{W}_{n+1}$ is injective, we have a unique $a_1(v) \in \mathcal{W}_2 \oplus \ldots \mathcal{W}_{n+1}$ so that $da_1(v) = \mathcal{I}(dv)$. By the isomorphism $\mathcal{H}^1(M) \otimes \mathbb{C} \cong H^1(M, \mathbb{C})$, we have a unique $a_2(v) \in \mathcal{W}_1$ so that
$$\phi_{n+1}(v) - \psi_{n+1}(a_1(v)) - \int_0^1 H(dv) - \psi_{n+1}(a_2(v))$$

is a exact form. Thus we take $a(v) = a_1(v) + a_2(v)$.
Let

$$b(v) = d^*G\left(\phi_{n+1}(v) - \psi_{n+1}(a(v)) - \int_0^1 H(dv)\right).$$

We extend $\mathcal{I} : \mathcal{M}_{n+1}^* \otimes \mathbb{C} \to \mathcal{N}_{n+1}^*$ and $H : \mathcal{M}_{n+1}^* \otimes \mathbb{C} \to A^*(M) \otimes \mathbb{C} \otimes$
(t, dt) so that $\mathcal{I}(v) = a(v)$ and

$$H(v) = \psi(\mathcal{I}(v)) + \int_0^t H(dv) + d(b(v)t).$$

This construction is well-defined.

Remark 5.1. We can easily check $\mathcal{I}(\mathcal{V}_2) \subset \mathcal{W}_2$ with the homotopy $H :$
$\mathcal{M}_2^* \otimes \mathbb{C} \to A^*(M) \otimes \mathbb{C} \otimes [t, dt]$ so that

$$H(v) = -2\partial\alpha(v) + d\alpha(v)t + \alpha(v)dt$$

for $v \in \mathcal{V}_2$ where $\alpha(v) = \sqrt{-1}F^c(\phi_1(dv))$. But for $n \geq 3$, $\mathcal{I}(\mathcal{V}_n) \not\subset \mathcal{W}_n$
may not hold and so $\mathcal{I}$ may not be compatible with the gradings $\mathcal{M}^* =$
$\bigoplus \mathcal{M}^*(n)$ and $\mathcal{N}^* = \bigoplus \mathcal{N}^*(n)$.

Theorem 5.2. *Define the filtration* $F^r(\mathcal{M}^* \otimes \mathbb{C}) = \mathcal{I}^{-1}(F^r(\mathcal{N}^*))$ *where*
$F^r(\mathcal{N}^*) = \bigoplus_{p \geq r} \mathcal{N}^*(p, q)$. *Then,* (W_*, F^*) *is an* $\mathbb{R}$-*mixed Hodge structure*
on $\mathcal{M}^*$ *such that the product, differential operator and the induced map*
$\phi : H^*(\mathcal{M}^*) \to H^*(M, \mathbb{R})$ *are morphisms of* $\mathbb{R}$-*mixed Hodge structures.*

Proof. This is in fact one of Morgan's mixed Hodge structures for the $\mathbb{R}$-mixed Hodge diagram as in Example 4.2. But, we give the explicit proof.

Forgetting the maps ϕ and ψ, for each $i \geq 2$, $\mathcal{V}_i$ and $\mathcal{W}_i$ are determined
by $\mathcal{V}_1$ with the cup-product $\mathcal{V}_1 \wedge \mathcal{V}_1 \to H^2(M, \mathbb{R})$ and $\mathcal{V}_1$ with the cup-product $\mathcal{W}_1 \wedge \mathcal{W}_1 \to H^2(M, \mathbb{C})$ respectively in the same manner. Since
$\mathcal{V}_1 \otimes \mathbb{C} = \mathcal{W}_1$, we have the identification $\mathcal{V}_i \otimes \mathbb{C} = \mathcal{W}_i$ and so $\mathcal{M}^* \otimes \mathbb{C} = \mathcal{N}^*$.
Consider $\mathcal{I}$ as a self-map on $\mathcal{M}^* \otimes \mathbb{C}$. Then, $\mathcal{I}$ preserves the filtration
W_* but we do not have $\mathcal{I}(\mathcal{M}^*(n) \otimes \mathbb{C}) \subset \mathcal{M}^*(n) \otimes \mathbb{C}$ for the grading
$\mathcal{M}^*(n) = \bigoplus \mathcal{M}^*(n)$. By the grading, we have $Gr^W(\mathcal{M}^*(n)) = \bigoplus \mathcal{M}^*(n) =$
$\mathcal{M}^*$ where $Gr^W = \bigoplus_i Gr_i^W$. We consider the induced map $Gr^W(\mathcal{I})$ on
$Gr^W(\mathcal{M}^*(n) \otimes \mathbb{C}) = \mathcal{M}^* \otimes \mathbb{C}$. We prove that $Gr^W(\mathcal{I})$ is the identity by
the induction. On $\mathcal{M}_1^1 \otimes \mathbb{C} = \bigwedge \mathcal{V}_1 \otimes \mathbb{C}$, this is obvious. We assume that
$Gr^W(\mathcal{I}) = \text{id}$ on $\mathcal{M}_n^1 \otimes \mathbb{C} = \bigwedge(\mathcal{V}_1 \oplus \cdots \oplus \mathcal{V}_n) \otimes \mathbb{C}$. Then, considering
$\mathcal{M}_{n+1}^1 \otimes \mathbb{C} = \mathcal{M}_n^1 \otimes \mathcal{V}_{n+1} \otimes \mathbb{C}$, for $v \in \mathcal{V}_{n+1}$, we have $\mathcal{I}(v) \in (\mathcal{V}_1 \oplus \cdots \oplus$
$\mathcal{V}_{n+1}) \otimes \mathbb{C}$ and $\mathcal{I}(dv) = d\mathcal{I}(v)$. By the inductive assumption, we have
$dv = dGr_W(\mathcal{I})(v)$. Since d is injective on $\mathcal{V}_{n+1}$, $v = Gr^W(\mathcal{I})(v)$. Thus,

we have proved that $Gr^W(\mathcal{I})$ is the identity. By this, we can easily see that the filtrations (W_*, F^*) induces a split $\mathbb{R}$-mixed Hodge structure on $Gr^W(\mathcal{M}^*)$ in the same way as Example 4.1. Hence, (W_*, F^*) is also an $\mathbb{R}$-mixed Hodge structure on $\mathcal{M}^*$. □

Remark 5.2. In general, on compact Kähler manifolds M such that $\mathcal{V}_n$ are non-trivial for $3 \geq n$, it is very difficult to compute our $\mathbb{R}$-mixed Hodge structures. Even if M are compact Riemann surfaces of genus ≥ 2 with hyperbolic metrics, it is too complicated to reach the complete computation.

By Torelli's theorem, compact Riemann surfaces are classified by polarized "$\mathbb{Z}$"-Hodge structures on the integral first cohomology. On the other hand, we can not completely distinguish compact Riemann surfaces of genus ≥ 2 by the $\mathbb{R}$-Hodge structures on $\mathcal{H}^1(M)$. It may be interesting to study whether our $\mathbb{R}$-mixed Hodge structures distinguish compact Riemann surfaces of genus ≥ 2 with hyperbolic metrics.

Let M_1 and M_2 be compact complex manifolds with a Kähler metrics g_1 and g_2. Take the canonical 1-minimal models $\phi_1 : \mathcal{M}_1^* \to A^*(M_1)$ and $\phi_2 : \mathcal{M}_2^* \to A^*(M_2)$ as in Construction 5.1 and $\psi_1 : \mathcal{N}_1^* \to A^*(M_1) \otimes \mathbb{C}$ and $\psi_2 : \mathcal{N}_2^* \to A^*(M_2) \otimes \mathbb{C}$ as in Construction 5.2.

Let $f : M_2 \to M_1$ be a holomorphic isometric (i.e. $g_2 = f^*g_1$) mapping. Then $f^* : A^*(M) \to A^*(M_2)$ commutes with all differential , co-differential, Green's operators as above. Hence, we have $f_{\mathcal{M}} : \mathcal{M}_1^* \to \mathcal{M}_2^*$ (resp. $f_{\mathcal{N}} : \mathcal{N}_1^* \to \mathcal{N}_2^*$) so that $\phi_2 \circ f_{\mathcal{M}} = f \circ \phi_1$ (resp. $\psi_2 \circ f_{\mathcal{N}} = f \circ \phi_1$) and $f_{\mathcal{M}}(\mathcal{M}_1^*(n)) \subset \mathcal{M}_2^*(n)$ (resp. $f_{\mathcal{N}}(\mathcal{N}_1^*(p,q)) \subset \mathcal{N}_2^*(p,q)$) for the grading (resp. bigrading). Take the isomorphisms $\mathcal{I}_1 : \mathcal{M}_1^* \otimes \mathbb{C} \to \mathcal{N}_1$ and $\mathcal{I}_1 : \mathcal{M}_2^* \otimes \mathbb{C} \to \mathcal{N}_2$ and the homotopies $H_1 : \mathcal{M}_1^* \otimes \mathbb{C} \to A^*(M_1) \otimes \mathbb{C} \otimes (t, dt)$ and $H_2 : \mathcal{M}_2^* \otimes \mathbb{C} \to A^*(M_2) \otimes \mathbb{C} \otimes [t, dt]$ as in Construction 5.3.

Proposition 5.2. $f_{\mathcal{M}} : \mathcal{M}_1^* \to \mathcal{M}_2^*$ is a morphism of $\mathbb{R}$-mixed Hodge structures.

Proof. It is sufficient to prove $\mathcal{I}_2 \circ f_{\mathcal{M}} = f_{\mathcal{N}} \circ \mathcal{I}_1$. We will inductively prove that for each steps $\phi_{1,n} : \mathcal{M}_{1,n}^* \to A^*(M) \otimes \mathbb{C}$, $\psi_{1,n} : \mathcal{N}_{1,n}^* \to A^*(M) \otimes \mathbb{C}$, $\phi_{2,n} : \mathcal{M}_{2,n}^* \to A^*(M) \otimes \mathbb{C}$ and $\psi_{2,n} : \mathcal{N}_{2,n}^* \to A^*(M) \otimes \mathbb{C}$ as in Constructions 5.1 and 5.2, we have $\mathcal{I}_2 \circ f_{\mathcal{M}} = f_{\mathcal{N}} \circ \mathcal{I}_1$ and $H_2 \circ f_{\mathcal{M}} = (f^* \otimes \mathrm{id}_{(t,dt)}) \circ H_1$. For $n = 1$, this claim is obvious. Assuming this claim for n, we prove for $n+1$. For $v \in \mathcal{V}_{n+1}$, we take a unique $a(v) \in \mathcal{N}_{1,n+1}^1$ so that $\mathcal{I}_1(dv) = da(v)$ and

$$\phi_{1,n+1}(v) - \psi_{1,n+1}(a(v)) - \int_0^1 H_1(dv)$$

is an exact form. Then, by the inductive assumption, we can say that $\mathcal{I}_2(df_{\mathcal{M}}(v)) = df_{\mathcal{N}}(a(v))$ and

$$\phi_{2,n+1}(f_{\mathcal{M}}(v)) - \psi_{2,n+1}(f_{\mathcal{N}}a(v)) - \int_0^1 H_2(df_{\mathcal{M}}(v))$$

is a exact form. Thus, we have $\mathcal{I}_2(f_{\mathcal{M}}(v)) = f_{\mathcal{N}}(a(v)) = f_{\mathcal{N}}(\mathcal{I}_1(v))$. Let

$$b(f_{\mathcal{M}}(v)) = d^*G\left(\phi_{2,n+1}(f_{\mathcal{M}}(v)) - \psi_{2,n+1}(f_{\mathcal{N}}a(v)) - \int_0^1 H_2(df_{\mathcal{M}}(v))\right).$$

Then, we have

$$H_2(f_{\mathcal{M}}(v)) = \psi_{2,n+1}(\mathcal{I}_2(f_{\mathcal{M}}(v))) + \int_0^t H_2(df_{\mathcal{M}}(v)) + d(b(f_{\mathcal{M}}(v))t).$$

By the inductive assumption and $\mathcal{I}_2(f_{\mathcal{M}}(v)) = f_{\mathcal{N}}(a(v)) = f_{\mathcal{N}}(\mathcal{I}_1(v))$, we can say that $b(f_{\mathcal{M}}(v)) = f^*(b(v))$ and

$$H_2(f_{\mathcal{M}}(v))$$
$$= f^*(\psi_{1,n+1}(\mathcal{I}_1(v)) + (f^* \otimes \mathrm{id}_{(t,dt)})\int_0^t H_1(dv) + (f^* \otimes \mathrm{id}_{(t,dt)})d(b(v)t)$$
$$= (f^* \otimes \mathrm{id}_{(t,dt)})(H_1(v)).$$

Hence the proposition follows. □

Remark 5.3. In general, a mixed Hodge structure on the 1-minimal model of $A^*(M)$ constructed by Morgan is not functorial for holomorphic maps between Kähler manifolds (see [10]). Thus, unlike the Hodge structures on the cohomology, Morgan's original mixed Hodge structures are not compatible with complex analytic geometry. Our canonical mixed Hodge structures are compatible with "Kähler geometry".

We note that in [6] Hain constructed the mixed Hodge structures on the $\mathbb{R}$-Malcev Lie algebra of the fundamental groups $\pi_1(M, x)$ of algebraic varieties M which are functorial with respect to morphisms of "pointed" complex algebraic varieties by using Chen's iterated integrals. In fact, they vary with the base points x.

Remark 5.4. Our canonical mixed Hodge structures can be applicable to constructing unipotent variations of mixed Hodge structure over compact Kähler manifolds ([7]).

In [7], we also give a way of constructing the non-unipotent variations of mixed Hodge structure.

Let M be a compact even-dimensional manifold. Suppose M admits a smooth family $\{J_t\}$ of complex structures on M and a smooth family $\{g_t\}$ of Riemannian metrics on M such that they are parametrized by a connected domain $B \subset \mathbb{R}^m$ and each g_t is a Kähler metric which is compatible with J_t. This situation is very considerable because of the Kodaira-Spencer stability theorem of Kähler manifolds [8]. We consider the canonical 1-minimal models $\phi_t : \mathcal{M}^* \to A^*(M)$ and $\psi_t : \mathcal{N}^* \to A^*(M) \otimes \mathbb{C}$ associated each Kähler manifold (M, J_t, g_t) as above. Then, since $\mathcal{V}_1, \mathcal{V}_2, \dots$ and $\mathcal{W}_1, \mathcal{W}_2, \dots$ with the differential operators are determined by the first cohomology of M and its cup-product, as filtered DGAs, $(\mathcal{M}^*, W_*)$ and $(\mathcal{N}^*, W_*)$ do not depend on the choices of a complex structure and a Kähler metric. Thus, we have the family of $\mathbb{R}$-mixed Hodge structures $\{(W_*, F_t^*)\}_{t \in B}$. For the isomorphism $\mathcal{I}_t$ with the homotopy H_t as in Construction 5.3, by $F_t^r(\mathcal{M}^* \otimes \mathbb{C}) = \mathcal{I}_t^{-1}(F_t^r(\mathcal{N}^*))$ and $F_t^r(\mathcal{N}^*) = \bigoplus_{p \geq r} \mathcal{N}_t^*(p, q)$, F_t^* depend on the family of isomorphisms $\{\mathcal{I}_t\}_{t \in B}$ and the bigradings $\mathcal{W}_n = \bigoplus_{p+q=n} \mathcal{W}_t^{p,q}$. We can see that $\mathcal{I}_t$ varies smoothly on B. For $v \in \mathcal{V}_{n+1}$ with $n \geq 1$, we have $\mathcal{I}_t(v) = a_1(v, t) + a_2(v, t)$ so that

- $a_1(v, t) \in \mathcal{W}_2 \oplus \dots \mathcal{W}_{n+1}$ and $da_1(v, t) = \mathcal{I}_t(dv)$
-

$$a_2(v, t) = P_t \left(\phi_{n+1}(v) - \psi_{n+1}(a_1(v)) - \int_0^1 H_t(dv) \right)$$

where P_t is the projection $A^1(M) \otimes \mathbb{C} \to \mathcal{H}_{g_t}^1$ for the Laplacian operator associated with g_t.

It is known that the Green operators and P_t depend smoothly on B see [8]. Thus, we can easily show that $\mathcal{I}_t$ varies smoothly on B by induction. The bigradings $\mathcal{W}_n = \bigoplus_{p+q=n} \mathcal{W}_t^{p,q}$ are determined by the Hodge structure $\mathcal{H}^1(M) \otimes \mathbb{C} = \mathcal{H}_t^{1,0}(M) \oplus \mathcal{H}_t^{0,1}(M)$ and the cup-product. The decomposition $\mathcal{H}^1(M) \otimes \mathbb{C} = \mathcal{H}_t^{1,0}(M) \oplus \mathcal{H}_t^{0,1}(M)$ is in fact given by the action of J_t on $\mathcal{H}^1(M)$. Thus, this decomposition depend smoothly on B. Since the cup-product is a topological invariant, we can say that the bigradings $\mathcal{W}_n = \bigoplus_{p+q=n} \mathcal{W}_t^{p,q}$ depend smoothly on B. Thus we can say that $\mathbb{R}$-mixed Hodge structures $\{(W_*, F_t^*)\}_{t \in B}$ on $\mathcal{M}^*$ depend smoothly on B.

Studies of the smooth family $\{(W_*, F_t^*)\}_{t \in B}$ of $\mathbb{R}$-mixed Hodge structures in terms of variation of (mixed) Hodge Structures and period mapping are left for a future work.

References

[1] J. Amorós, M. Burger, K. Corlette, D. Kotschick, and D. Toledo, *Fundamental groups of compact Kähler manifolds*, Mathematical Surveys and Monographs, vol. 44, A.M.S., Providence, RI, 1996.

[2] E. Cattani, A. Kaplan and W. Schmid, Degeneration of Hodge structures, *Ann. of Math.*, (2) **123**, 457–535 (1986).

[3] P. Deligne, Théorie de Hodge. II. *I.H.E.S. Publ. Math.*, **40**, 5–57 (1971).

[4] P. Deligne, P. Griffiths, J. Morgan, and D. Sullivan, Real homotopy theory of Kahler manifolds, *Invent. Math.*, **29**, no. 3, 245–274 (1975).

[5] P. Griffiths, J. Morgan, *Rational homotopy theory and differential forms*, Second ed., Progress in Math., **16**. Springer, New York, 2013.

[6] R. M. Hain, The de Rham homotopy theory of complex algebraic varieties. I, *K-Theory* **1**, no. 3, 271–324 (1987).

[7] H. Kasuya, Techniques of constructions of variations of mixed Hodge structures. Preprint (2016)

[8] K. Kodaira, D. C. Spencer, On deformations of complex analytic structures. III. Stability theorems for complex structures. *Ann. of Math.* (2) **71**, 43–76 (1960).

[9] J. W. Morgan, The algebraic topology of smooth algebraic varieties. *I.H.E.S. Publ. Math.* **48**, 137–204 (1978).

[10] J. W. Morgan, Correction to "The algebraic topology of smooth algebraic varieties". *I.H.E.S. Publ. Math.* **64**, 185 (1986).

[11] C. A. M. Peters and J. H. M. Steenbrink, *Mixed Hodge structures*, Ergebnisse der Mathe- matik und ihrer Grenzgebiete, 3. Fogle, A Series of Modern Surveys in Mathematics, **52**, Springer, 2008.

[12] D. Sullivan, Infinitesimal computations in topology. *I.H.E.S. Publ. Math.* **47**, 269–331 (1977–78).

[13] C. Voisin, *Hodge theory and complex algebraic geometry. I*, Cambridge Studies in Advanced Mathematics, **76**, Cambridge University Press, Cambridge, 2002.

Received December 19, 2016

Contemporary Perspectives
in Differential Geometry
and its Related Fields   53 – 66

F-GEODESICS ON THE COTANGENT BUNDLE
OF A WEYL MANIFOLD

100 years since the birth of Professor Tominosuke Otsuki (1917–2011)

Cornelia-Livia BEJAN

*Department of Mathematics, "Gh. Asachi" Technical University,
Bd. Carol I, no.11, corp A, 700506 Iasi, Romania*

** Seminarul Matematic, Universitatea "Alexandru Ioan Cuza",
Bd. Carol I, no.11, 700506 Iasi, Romania
E-mail: bejanliv@yahoo.com
URL: http://math.etc.tuiasi.ro/bejan/*

İlhan GÜL

*Department of Mathematics, Istanbul Technical University,
34469 Maslak, Istanbul, Turkey
E-mail: igul@itu.edu.tr*

We introduce here the notion of distorted Riemann extension as a generalization of classical Riemann extension. We construct a class of semi-Riemannian
metrics of neutral signature on the total space of the cotangent bundle of a manifold endowed with Weyl structure and we obtain an equivariance result. Then
we provide a class of F-geodesics (which generalize both classical geodesics and
magnetic curves) on T^*M, with respect to this distorted Riemann extensions.

Keywords: Cotangent bundle; Riemann extension; Weyl structure; F-geodesics.

1. Introduction

The geometry of the tangent bundle of a manifold is a rich context in
which several research topics can be developed. Different from the huge
literature devoted to the tangent bundle, the geometry of the cotangent
bundle is less well represented. The framework of our study is the total
space of the cotangent bundle T^*M of a manifold (M, ∇) endowed with
an affine symmetric connection. On T^*M (named in physics the phase
space), Patterson and Walker ([9]) and then Willmore ([11]) introduced the
Riemann extension as a semi-Riemannian metric of neutral signature. Since
then, several extensions of this (classical) Riemann extension were provided
in literature. We mention here only the natural Riemann extension given
by Kowalski and Sekizawa (see [10], [8]) and applied recently in [3]. As

---

*Postal address.

a generalization of the Riemann extension, we construct here a family of metrics (called distorted Riemann extensions) of neutral signature which depend on arbitrary positive constants and arbitrary smooth functions on the base manifold. For these distorted Riemann extensions we prove an equivariance result (Theorem 2.1). Here we recall that H. Weyl introduced in 1918 the notion of Weyl structure to formulate a unified field theory. A Weyl structure on a manifold M is defined as a torsion-free connection D (called a Weyl connection) and a conformal class $[g]$ of metrics, which is preserved by this connection. Further, we deal with a Weyl structure in Proposition 3.1, to obtain the behavior of our distorted Riemann extensions under the gauge transformations of the metrics in the conformal class $[g]$. Once one has a linear connection on a manifold M, then special curves as geodesics or their generalization can be studied.

Many papers on magnetic curves, inspired from theoretical physics have appeared, such as the Lorentz force, the electro-magnetic tensor field, as well as some special forces involved in the Euler-Lagrange equations from Lagrangian mechanics. All these curves can be studied in a unitary way, if one considers an arbitrary $(1,1)$-tensor field F on a manifold M. By using it, one may deal with a notion which generalizes both the classical equations of geodesics and magnetic curves, namely the F-geodesics on manifolds, introduced in [2]. In the present work, based on Proposition 4.2 we obtain in Corollary 4.1 and Corollary 4.2 some classes of geodesics and respectively F-geodesics with respect to the distorted Riemann extensions associated to a Weyl connection.

2. Geometric objects on cotangent bundle

We recall here some basic facts on the geometry of cotangent bundles. If M is a manifold and T^*M is its cotangent bundle, then the natural projection $\pi : T^*M \longmapsto M$ is defined by $\pi(x,\omega) = x$ for any point $(x,\omega) \in T^*M$, where $x \in M$ and ω is a one-form on T_xM. On T^*M, a local coordinate system around $(x_0, \omega_0) \in T^*M$ is given by $(\pi^{-1}(U); x^1, ..., x^n, x^{1*}, ..., x^{n*})$, where $(U; x^1, ..., x^n)$ denotes a local coordinate system on M around $x_0 \in M$, the function $x^i \circ \pi$ on $\pi^{-1}(U)$ is identified with x^i on U, and the notation $x^{i*} = \omega_i = \omega((\partial/\partial x^i)_x)$ is used at any point $(x, w) \in \pi^{-1}(U)$, for any $i = 1, ..., n$. If we denote $\partial_i = \partial/\partial x^i$ and $\partial_{i*} = \partial/\partial \omega_i$, $i = 1, ..., n$, then at each point $(x, \omega) \in T^*M$, it follows that

$$\{(\partial_1)_{(x,\omega)}, ..., (\partial_n)_{(x,\omega)}, (\partial_{1*})_{(x,\omega)}, ..., (\partial_{n*})_{(x,\omega)}\} \tag{1}$$

is a basis for the tangent space $T_{(x,\omega)}T^*M$.

Let $\mathcal{F}(M)$ be the set of all smooth functions on M. Then any function on M can be lifted to a function on T^*M. That is, if $f \in \mathcal{F}(M)$, then its vertical lift f^v on T^*M is defined by $f^v = f \circ \pi$.

Let $\chi(M)$ be the set of all smooth vector fields on M. For any $X \in \chi(M)$, the evaluation map X^v of X is a smooth function on T^*M defined by

$$X^v(x, \omega) = \omega(X_x). \tag{2}$$

The importance of the evaluation map is shown in the following characterization of any vector field on T^*M by its action on evaluation maps:

Proposition 2.1 ([12]). *Let U and V be vector fields on T^*M. If $U(Z^v)$ $= V(Z^v)$ holds for all $Z \in \chi(M)$, then $U = V$.*

Hence, the complete lift of any vector field X on M is a vector field X^c on T^*M defined by

$$X^c(Z^v) = [X, Z]^v \quad \text{for each } Z \in \chi(M). \tag{3}$$

Its expression in local coordinates at any point $(x, \omega) \in T^*M$ is

$$X^c_{(x,\omega)} = \sum_{i=1}^{n} \xi^i(x)(\partial_i)_{(x,\omega)} - \sum_{h,i=1}^{n} \omega_h(\partial_i \xi^h)(x)(\partial_{i*})_{(x,\omega)}, \tag{4}$$

where

$$X = \sum_{i=1}^{n} \xi^i \partial_i.$$

Therefore we see $X^c(f^v) = (Xf)^v$ for arbitrary $f \in \mathcal{F}(M)$.

The vertical lift of a one-form α on M is a vector field α^v tangent to T^*M, which can be defined (based on Proposition 2.1) only by its action on evaluation maps as

$$\alpha^v(Z^v) = (\alpha(Z))^v \quad \text{for each } Z \in \chi(M). \tag{5}$$

Remark 2.1. We note the following:

(i) Any vector field $U \in \chi(T^*M)$ decomposes into the complete and the vertical parts, that is, there exist a vector field $Z \in \chi(M)$ and a one form $\theta \in \Omega^1(M)$ on M such that

$$U = Z^c + \theta^v;$$

(ii) Any vector field $U \in \chi(T^*M)$ is identically zero if and only if both its complete and vertical parts (i.e. Z^c and θ^v) vanish identically on T^*M;

(iii) Any vector field $Z \in \chi(M)$ is identically zero if and only if its complete lift $Z^c \in \chi(T^*M)$ vanishes identically.

The vertical vector field L of Liouville type on T^*M is a global vector field defined in terms of local coordinate systems by

$$L = \sum_{i=1}^{n} \omega_i \partial_{i*}. \tag{6}$$

On T^*M the Lie bracket satisfies

$$[X^c, Y^c] = [X, Y]^c, \quad [X^c, \alpha^v] = (\mathcal{L}_X \alpha)^v, \tag{7}$$

$$[\alpha^v, \beta^v] = 0 = [X^c, L], \quad [\alpha^v, L] = \alpha^v, \tag{8}$$

for arbitrary $X, Y \in \chi(M)$, $\alpha, \beta \in \Omega^1(M)$, where $\mathcal{L}_X$ denotes the Lie derivative with respect to X.

If F is a $(1,1)$-tensor field on M, then its complete lift F^c is a $(1,1)$-tensor field on T^*M defined by

$$\begin{aligned} F^c \alpha^v &= (\alpha \circ F)^v, \\ F^c X^c &= (FX)^c + (\mathcal{L}_X F)^v, \end{aligned} \tag{9}$$

for arbitrary $X \in \chi(M), \alpha \in \Omega^1(M)$.

Remark 2.2. All formulas and all lifts defined above still hold if the geometric objects such as f, α, X, Z are defined only locally.

We stress here that any tensor field of type $(0, s)$ on T^*M is determined by its action on complete lifts of vector fields on M (see [12]). As for instance, the Riemann extension is a metric on the cotangent manifold which is defined as follows:

Definition 2.1. [9, 11] Associated to any torsion-free connection ∇ on a manifold M, its Riemann extension is the neutral signature metric on T^*M defined by

$$G_{(x,\omega)}(X^c, Y^c) = -\omega(\nabla_{X_x} Y + \nabla_{Y_x} X), \tag{10}$$

$$G_{(x,\omega)}(X^c, \alpha^v) = \alpha_x(X_x), \tag{11}$$

$$G_{(x,\omega)}(\alpha^v, \beta^v) = 0. \tag{12}$$

The classical Riemann extension was used in mathematical literature in relation to the Osserman problem, Walker manifolds, almost para-Hermitian manifolds, non-Lorentzian geometry and so on.

Next, we generalize the classical Riemann extension G to the distorted Riemann extension $G + S$ by adding to G a distorted tensor field S which is a symmetric $(0, 2)$-tensor field on T^*M. The particular case when $S = \pi^* Q$, (for any symmetric $(0, 2)$-tensor field Q on M), is well known in literature, under the name of "deformed Riemann extension". As it was pointed out in [4], these deformed Riemann extensions are typical examples of Walker metric since $\ker \pi_*$ is a parallel degenerate plane field of maximum rank on T^*M. Another particular case is the so-called natural Riemann extension. It is obtained when S is defined at any point $(x, \omega) \in T^*M$ by $S = b\omega \otimes \omega$, where b is a real constant. The natural Riemann extensions were described first in [10] and then studied by Kowalski and Sekizawa in [7, 8]. For the concept of naturality, we refer to [6]. We note that the natural Riemann extension is a distorted Riemann extension but it is not a deformed Riemann extension, since $\omega \otimes \omega$ is not the pull back $\pi^* Q$ of a symmetric $(0, 2)$-tensor field Q on M.

Definition 2.2. Let (M, ∇) be a manifold endowed with a torsion-free affine connection. Then for any smooth real function $f \in \mathcal{F}(M)$ and any positive constant a, we can define a semi-Riemannian metric $\mathcal{G}^{(\nabla, f, a)}$ on the total space of T^*M by

$$
\begin{aligned}
\mathcal{G}^{(\nabla, f, a)}(X^c, Y^c)_{(x, \omega)} &= -a\omega(\nabla_{X_x} Y + \nabla_{Y_x} X) + 2df(X_x)\omega(Y_x) \\
&\quad + 2df(Y_x)\omega(X_x), \\
\mathcal{G}^{(\nabla, f, a)}(\alpha^v, X^c)_{(x, \omega)} &= a\alpha(X_x), \\
\mathcal{G}^{(\nabla, f, a)}(\alpha^v, \beta^v) &= 0,
\end{aligned}
\tag{13}
$$

for arbitrary $X, Y \in \chi(M)$, $\alpha, \beta \in \Omega^1(M)$. Hence $\mathcal{G}^{(\nabla, f, a)}$ is a distorted Riemann extension for which S is defined at any point $(x, \omega) \in T^*M$ by $S = df \otimes \omega + \omega \otimes df$.

Like the case of the classical Riemann extension $\mathcal{G}^{(\nabla, 0, 1)}$ (obtained for $f = 0$ and $a = 1$), for any smooth real function $f \in \mathcal{F}(M)$, the distorted Riemann extension $\mathcal{G}^{(\nabla, f, a)}$ is a semi-Riemannian metric of neutral signature.

If ϕ denotes a (local) diffeomorphism of M, then its lift Φ to T^*M was defined in [7] by

$$
\Phi(x, \omega_x) = (\phi(x), (\phi^{-1})^*(\omega_x))
\tag{14}
$$

for all $(x, \omega_x) \in T^*M$, where for the sake of simplicity, we denote $(x, w) \in T^*M$ instead of $(x, \omega_x) \in T^*M$, when it is convenient. This lift satisfies the following properties:

Proposition 2.2 ([7]). *For arbitrary* $X \in \chi(M)$, $\alpha \in \Omega^1(M)$, *we have*

$$X^v \circ \Phi = ((\phi^{-1})_* X)^v, \tag{15}$$

$$\Phi_*(X^c) = (\phi_* X)^c, \tag{16}$$

$$\Phi_*(\alpha^v) = ((\phi^{-1})^* \alpha)^v. \tag{17}$$

Theorem 2.1. *Let* ϕ *be a (local) affine diffeomorphism of a manifold* M *with a symmetric affine connection* ∇. *Then for any smooth real function* $f \in \mathcal{F}(M)$ *and any positive constant* a, *the lift* Φ *of* ϕ

$$\Phi : (T^*M, \mathcal{G}^{(\nabla, f \circ \phi, a)}) \longrightarrow (T^*M, \mathcal{G}^{(\nabla, f, a)}), \tag{18}$$

defined by (14) *is a (local) isometry between semi-Riemannian manifolds.*

Proof. We have to show the following relation for every $U, V \in \chi(T^*M)$:

$$\mathcal{G}^{(\nabla, f, a)}(\Phi_*(U), \Phi_*(V)) = \mathcal{G}^{(\nabla, f \circ \phi, a)}(U, V). \tag{19}$$

Since ϕ is a local diffeomorphism of (M, ∇), it satisfies

$$\phi_*(\nabla_X Y) = \nabla_{\phi_* X}(\phi_* Y), \tag{20}$$

for arbitrary $X, Y \in \chi(M)$. In what follows, we prove the relation (19) by replacing the pair (U, V) consecutively by (X^c, Y^c), (X^c, α^v) and (α^v, β^v). First, we show that at each point $(x, \omega_x) \in T^*M$

$$\mathcal{G}^{(\nabla, f, a)}(\Phi_*(X^c), \Phi_*(Y^c)) = \mathcal{G}^{(\nabla, f \circ \phi, a)}(X^c, Y^c), \tag{21}$$

for arbitrary $X, Y \in \chi(M)$. For this purpose, we use (16) and the first formula of (13) in the following computation:

$$\begin{aligned}
&\left[\mathcal{G}^{(\nabla, f, a)}\big(\Phi_*(X^c), \Phi_*(Y^c)\big) \right]_{\Phi(x, \omega_x)} \\
&= \left[\mathcal{G}^{(\nabla, f, a)}\big((\phi_* X)^c, (\phi_* Y)^c\big) \right]_{(\phi(x), (\phi^{-1})^*(\omega_x))} \\
&= -a\left[(\phi^{-1})^*(\omega_x)\big(\big(\nabla_{\phi_* X}(\phi_* Y) + \nabla_{\phi_* Y}(\phi_* X)\big)_{\phi(x)}\big) \right] \\
&\quad + df\big((\phi_* X)_{\phi(x)}\big)\,\big((\phi^{-1})^*(\omega_x)\big)\big((\phi_* Y)_{\phi(x)}\big) \\
&\quad + \big((\phi^{-1})^*(\omega_x)\big)\big((\phi_* X)_{\phi(x)}\big)\,df\big((\phi_* Y)_{\phi(x)}\big) \\
&= -a\left[\big((\phi^{-1})^*(\omega_x)\big)\big(\phi_{*x}((\nabla_X Y + \nabla_Y X)_x)\big) \right] \\
&\quad + df\big(\phi_{*x}(X_x)\big)\,\big((\phi^{-1})^*(\omega_x)\big)(\phi_{*x}(Y_x)) \\
&\quad + \big((\phi^{-1})^*(\omega_x)\big)(\phi_{*x}(X_x))\,df\big(\phi_{*x}(Y_x)\big) \\
&= -a\omega_x\big((\nabla_X Y + \nabla_Y X)_x\big) \\
&\quad + d(f \circ \phi)(X_x)\,\omega_x(Y_x) + \omega_x(X_x)\,d(f \circ \phi)(Y_x) \\
&= \left[\mathcal{G}^{(\nabla, f \circ \phi, a)}(X^c, Y^c) \right]_{(x, \omega_x)}.
\end{aligned} \tag{22}$$

Next we show at each point $(x, \omega_x) \in T^*M$

$$\mathcal{G}^{(\nabla, f, a)}(\Phi_*(X^c), \Phi_*(\alpha^v)) = \mathcal{G}^{(\nabla, f \circ \phi, a)}(X^c, \alpha^v), \tag{23}$$

for arbitrary $X \in \chi(M)$, $\alpha \in \Omega^1(M)$. We use (16), (17) and the second formula of (13) in the following computation:

$$
\begin{aligned}
\left[\mathcal{G}^{(\nabla, f, a)}\big(\Phi_*(X^c), \Phi_*(\alpha^v)\big)\right]_{\Phi(x, \omega_x)} \\
= \left[\mathcal{G}^{(\nabla, f, a)}\big((\phi_* X)^c, (\phi_* \alpha)^v\big)\right]_{(\phi(x), (\phi^{-1})^*(\omega_x))} \\
= a\big((\phi^{-1})^* \alpha\big)_{\phi(x)}\big((\phi_* X)_{\phi(x)}\big) \\
= a\big((\phi^{-1})^* \alpha_x\big)(\phi_{*x}(X_x)) \\
= a\alpha_x(X_x) \\
= \left[\mathcal{G}^{(\nabla, f \circ \phi, a)}(X^c, \alpha^v)\right]_{(x, \omega_x)}.
\end{aligned}
\tag{24}
$$

The third formula of (13) shows that (19) holds if we replace (U, V) by (α^v, β^v). This relation together with (21) and (23) yield (19), which complete the proof. □

Remark 2.3.

(1) Different from the case of natural Riemann extension [7], where the natural Riemann extension $\breve{g}$ is invariant by Φ (i.e. $\Phi^* \breve{g} = \breve{g}$), in our case, Theorem 2.1 states that the distorted Riemann extension $\mathcal{G}^{(\nabla, f, a)}$ on T^*M remains in the same class of distorted Riemann extensions under the action of Φ, namely one has

$$\Phi^* \mathcal{G}^{(\nabla, f, a)} = \mathcal{G}^{(\nabla, f \circ \phi, a)}. \tag{25}$$

(2) If we take $f = 0$ and $a = 1$ in the relation (25), then we reobtain that Φ is an isometry of T^*M with respect to the (classical) Riemann extension.

3. Weyl connection

A Weyl structure W on a manifold M consists of a Riemannian metric g and a 1-form φ on M (see [5]). The pair (M, W) is called a Weyl manifold. Like the Riemannian case, there exists a unique torsion-free affine connection D on M, called the Weyl connection of W, such that $Dg = -2\varphi \otimes g$. Since the Weyl connection is required to be invariant under the gauge transformation

$$g \longmapsto \bar{g} = e^{2\lambda} g, \tag{26}$$

the 1-form φ must change as follows:

$$\varphi \longmapsto \bar{\varphi} = \varphi - d\lambda. \tag{27}$$

Therefore the conformal class $C(W)$ determined by g forms a primary underlying structure.

Let g be the original metric of a Weyl structure. Then its Levi-Civita connection ∇ is related to the Weyl connection D by

$$D_X Y = \nabla_X Y + \varphi(Y)X + \varphi(X)Y - g(X,Y)\xi, \tag{28}$$

for $X, Y \in \chi(M)$, where ξ is the dual vector field of φ with respect to g.

Remark 3.1. The invariance of the Weyl connection assures that the relation (28) holds if g, ∇, φ, ξ are replaced respectively by $\bar{g}$ (from (26)), $\bar{\nabla}$ (the Levi-Civita connection of $\bar{g}$), $\bar{\varphi}$ (from (27)) and $\bar{\xi}$ (the dual of $\bar{\varphi}$).

Under the above considerations, by using the evaluation map from (2), we obtain the following:

Proposition 3.1. *Let (M, W) be a Weyl manifold, $f \subset \mathcal{F}(M)$ and $a \in \mathbb{R}^*_+$. Let $\mathcal{G}^{(\bar{\nabla},f,a)}$ be the distorted Riemann extension defined as in (13), where $\bar{\nabla}$ denotes the Levi-Civita connection of any metric $\bar{g} \in C(W)$. Then, under a gauge transformation*

$$\bar{g} \longmapsto \bar{\bar{g}} = e^{2\lambda}\bar{g},$$

the distorted metric $\mathcal{G}^{(\bar{\nabla},f,a)}$ changes to

$$\mathcal{G}^{(\bar{\nabla},f,a)} \longmapsto \mathcal{G}^{(\bar{\bar{\nabla}},f+d\lambda,a)} - 2(grad\lambda)^v \pi^* \bar{g},$$

where $\bar{\bar{\nabla}}$ denotes the Levi-Civita connection of $\bar{\bar{g}}$.

4. F-geodesics

We recall from [2] a notion which generalizes and unifies both the classical geodesics and magnetic curves, namely the F-geodesics on manifolds. When a manifold M is endowed with a torsion free connection ∇ and a $(1,1)$-tensor field F, a smooth curve $\gamma : I \longmapsto M$ on M is said to be an F-geodesic if it satisfies

$$\nabla_{\dot{\gamma}(u)}\dot{\gamma}(u) = F\dot{\gamma}(u). \tag{29}$$

Remark 4.1.

(1) In particular, if F is identically zero, then (29) gives the notion of geodesic.

(2) When M is 3-dimensional and F is given by

$$FX = B \times X, \tag{30}$$

for each $X \in \Gamma(TM)$ with a magnetic induction B, an F-geodesic is called a normal magnetic curve.

(3) When M is a complex manifold with the complex structure J, if we set $F = \kappa J$ with a real constant κ, then an F-geodesic is a trajectory for a Kähler magnetic field (see [1]).

(4) In [2], one can find several classes of examples of F-geodesics which motivate their study.

From the relation (28), one obtains the following:

Proposition 4.1. *Let γ be a smooth curve $\gamma : I \to M$ on a Weyl manifold (M, W). Then γ can not be simultaneously a geodesic for the Weyl connection D and the Levi-Civita connection $\tilde{\nabla}$ of a metric $\tilde{g} \in C(W)$.*

Notation 4.1. Let T be a $(1, 1)$-tensor field on a manifold M.

(1) For any 1-form α on M, one can define $i_\alpha(T)$ as a 1-form on M by

$$i_\alpha(T)(X) = \alpha(TX) \tag{31}$$

for all $X \in \chi(M)$. For instance, for any $Y \in \chi(M)$, the $(1, 1)$-tensor field T is particularized by ∇Y, that can be seen as an endomorphism defined by

$$\chi(M) \ni Z \longmapsto \nabla_Z Y \in \chi(M). \tag{32}$$

(2) At any point $(x, \omega) \in T^*M$, the contracted vector field $C_\omega(T) \in \chi(T^*M)$ is defined (based on Proposition 2.1) only by its action on evaluation maps as

$$C_\omega(T)(X^v)_{(x,\omega)} = (TX)^v_{(x,\omega)} = \omega((TX)_x) \tag{33}$$

for all $X \in \chi(M)$.

Proposition 4.2. *Let (M, ∇) be a manifold endowed with a torsion-free affine connection whose curvature tensor field is denoted by R. Then for any smooth real function $f \in \mathcal{F}(M)$ and any positive constant a, it follows that the Levi-Civita connection $\nabla^{\mathcal{G}}$ of the distorted Riemann extension $\mathcal{G}^{(\nabla, f, a)}$ can be expressed on the total space of T^*M by*

$$
\begin{aligned}
(\nabla^{\mathcal{G}}_{X^c} Y^c)_{(x,\omega)} &= (\nabla_X Y)^c_{(x,\omega)} + C_\omega((\nabla X)(\nabla Y) + (\nabla Y)(\nabla X))_{(x,\omega)} \\
&\quad + C_\omega\big(R_x(., X)Y + R_x(., Y)X\big)_{(x,\omega)} \\
&\quad - (1/a)\big\{df(Y)X^c + df(X)Y^c\big\}_{(x,\omega)} \\
&\quad + \big\{\sigma(X, Y) + \sigma(Y, X)\big\}_{(x,\omega)},
\end{aligned}
\tag{34}
$$

$$(\nabla^{\mathcal{G}}_{X^c}\beta^v)_{(x,\omega)} = (\nabla_X\beta)^v_{(x,\omega)} + \frac{1}{a}\{df(X)\beta^v + \beta(X)(df)^v\}_{(x,\omega)}, \qquad (35)$$

$$(\nabla^{\mathcal{G}}_{\alpha^v}Y^c)_{(x,\omega)} = -(i_\alpha(\nabla Y))^v_{(x,\omega)} + \frac{1}{a}\{df(Y)\alpha^v + \alpha(Y)(df)^v\}_{(x,\omega)}, (36)$$

$$(\nabla^{\mathcal{G}}_{\alpha^v}\beta^v)_{(x,\omega)} = 0, \qquad (37)$$

for arbitrary $X, Y \in \chi(M)$, $\alpha, \beta \in \Omega^1(M)$, where

$$\sigma(X,Y) = -\frac{1}{a}\{2df(Y)C_\omega(\nabla X) + \omega(Y)(i_{df}(\nabla_X\cdot) - i_{df}(\nabla X))^v$$
$$+ \omega(\nabla_X Y)(df)^v + df(\nabla_X Y)L\} \qquad (38)$$
$$+ \frac{2}{a^2}\{df(X)df(Y)L + df(X)\omega(Y)(df)^v\}.$$

Proof. The expression for the Levi-Civita connection of a natural Riemann extension was obtained in [7]. That expression was used in [3] for the study of harmonicity. We apply here the same techniques as in [7] to obtain the relations (34)-(37). In what follows, we use the definition of the Levi-Civita connection $\nabla^{\mathcal{G}}$ of the distorted Riemann extension $\mathcal{G}^{(\nabla,f,a)}$ which is given as

$$2\mathcal{G}^{(\nabla,f,a)}(\nabla^{\mathcal{G}}_U V, H)$$
$$= U\mathcal{G}^{(\nabla,f,a)}(V,H) + V\mathcal{G}^{(\nabla,f,a)}(H,U) - H\mathcal{G}^{(\nabla,f,a)}(U,V) \qquad (39)$$
$$+ \mathcal{G}^{(\nabla,f,a)}([U,V],H) + \mathcal{G}^{(\nabla,f,a)}([H,U],V) - \mathcal{G}^{(\nabla,f,a)}([V,H],U),$$

for arbitrary $U, V, H \in \chi(T^*M)$. Next, we shall obtain the relations (34)–(37) from (39) by using the fact that $\mathcal{G}^{(\nabla,f,a)}$ is non degenerate, that is, the condition that the equality $\mathcal{G}^{(\nabla,f,a)}(U,V) = \mathcal{G}^{(\nabla,f,a)}(U,H)$ holds for all $U \in \chi(T^*M)$ implies $V = H$. To prove each of the relations (34)–(37), we also apply the formulas (8) and (13). Given arbitrary $\alpha, \beta \in \Omega^1(M)$ and $X \in \chi(M)$ we set $U = \alpha^v$, $V = \beta^v$ and $H = X^c$ in (39). By using (8) and (13), we find $\mathcal{G}^{(\nabla,f,a)}(\nabla^{\mathcal{G}}_{\alpha^v}\beta^v, X^c) = 0$. Similarly, for an arbitrary $\delta \in \Omega^1(M)$, by setting $U = \alpha^v$, $V = \beta^v$ and $H = \delta^v$ in (39), we have $\mathcal{G}^{(\nabla,f,a)}(\nabla^{\mathcal{G}}_{\alpha^v}\beta^v, \delta^v) = 0$. Thus we obtain $\nabla^{\mathcal{G}}_{\alpha^v}\beta^v = 0$. In case when H is taken to be the complete lift of vector fields, we use in computation the Notation 4.1 (1).

From the second relation of (13), we obtain (36). The relation (35) follows in a similar way, by replacing U, V in the relation (39) respectively by X^c, β^v for arbitrary $X \in \chi(M)$, $\beta \in \Omega^1(M)$. To obtain (34), we use again (39) in which each vector field U, V on T^*M is replaced by the complete lift of vector fields on M and after that, H is replaced consecutively by the complete lift of vector fields and by the vertical lift of one forms on M. Then we use three times the notation (33).

Firstly, we use (33) when the $(1,1)$-tensor field T is particularized by $R(\cdot, X)Y$ that can be seen as a $(1,1)$-tensor field defined by

$$\chi(M) \ni Z \longmapsto R(Z, X)Y \in \chi(M). \tag{40}$$

Secondly, we use (33) when the $(1,1)$-tensor field T is particularized by $(\nabla X)(\nabla Y)$ that can be seen as a $(1,1)$-tensor field defined by

$$\chi(M) \ni Z \longmapsto [(\nabla X)(\nabla Y)](Z) = (\nabla X)(\nabla_Z Y) = \nabla_{\nabla_Z Y} X \in \chi(M). \tag{41}$$

If we subtract from the vertical part of $\nabla^{\mathcal{G}}_{X^c} Y^c$ the terms containing the curvature R (see (40)) and the terms containing both $(\nabla X)(\nabla Y)$ and $(\nabla Y)(\nabla X)$ (see (41)), then we obtain an expression (which is symmetric in X and Y) that we denote by $\sigma(X, Y) + \sigma(Y, X)$.

Lastly, we use (33) in the expression of σ, where we also used (31). Summarizing up, we obtain

$$2\mathcal{G}^{(\nabla, f, a)}\big(\sigma(X, Y), Z^c\big)$$

$$= 2\mathcal{G}^{(\nabla, f, a)} \Big(-\frac{1}{a} \big[2df(Y) \, C_\omega(\nabla X) + \omega(\nabla_X Y)(df)^v + df(\nabla_X Y)L \big], Z^c \Big)$$

$$+ \frac{2}{a^2} \mathcal{G}^{(\nabla, f, a)} \big(2df(X)df(Y)L + 2df(X)\omega(Y)(df)^v, \, Z^c \big)$$

$$+ 2df\big(\nabla_Z X - \nabla_X Z\big) \omega(Y)$$

$$= 2\mathcal{G}^{(\nabla, f, a)} \Big(-\frac{1}{a} \big[2df(Y) \, C_\omega(\nabla X) + \omega(\nabla_X Y)(df)^v + df(\nabla_X Y)L \big], Z^c \Big)$$

$$+ \frac{2}{a^2} \mathcal{G}^{(\nabla, f, a)} \big(2df(X)df(Y)L + 2df(X) \, \omega(Y)(df)^v, \, Z^c \big)$$

$$+ 2\mathcal{G}^{(\nabla, f, a)} \Big(\frac{1}{a} \big(i_{df}(\nabla X) \big)^v \omega(Y), \, Z^c \Big) - 2df(\nabla_X Z) \omega(Y).$$

After a standard calculation we obtain (34). Therefore, we complete the proof. □

From the relation (37), we deduce the following:

Corollary 4.1. *Under the notations of Proposition 4.2, any integral curve of each vector field $\alpha^v \in T^*M$ (expressed as the vertical lift of any one-form α on M) is a (classical) geodesic on T^*M with respect to the metric $\mathcal{G}^{(\nabla, f, a)}$.*

Another consequence of Proposition 4.2 is the following:

64   C.-L. BEJAN & İ. GÜL

Corollary 4.2. *Let (M, W) be a Weyl manifold with the Weyl connection D and a $(1,1)$-tensor field F. Then the integral curves of a vector field $X \in \chi(M)$ on M are classical geodesics (resp. F-geodesics) with respect to D, provided any integral curve of X^c is a classical geodesic (resp. an F^c-geodesic) with respect to the metric $\mathcal{G}^{(D,f,a)}$ on T^*M, where a is a positive constant and f has zero covariant derivative with respect to X.*

Proof. Since the Weyl connection D is torsion free, for any smooth function f on M and any positive constant a, just like $\mathcal{G}^{(\nabla,f,a)}$, the distorted Riemann extension $\mathcal{G}^{(D,f,a)}$ on T^*M is defined by

$$
\begin{aligned}
&\mathcal{G}^{(D,f,a)}(X^c, Y^c)_{(x,\omega)} \\
&= -a\omega(D_{X_x}Y + D_{Y_x}X) + 2df(X_x)\omega(Y_x) + 2df(Y_x)\omega(X_x), \\
&\mathcal{G}^{(D,f,a)}(\alpha^v, X^c)_{(x,\omega)} = a\alpha(X_x), \\
&\mathcal{G}^{(D,f,a)}(\alpha^v, \beta^v) = 0,
\end{aligned} \tag{42}
$$

for arbitrary $X, Y \in \chi(M)$, $\alpha, \beta \in \Omega^1(M)$. The Levi-Civita connection $\mathscr{D}$ of $\mathcal{G}^{(D,f,a)}$ is given as for $\nabla^{\mathcal{G}}$ of $\mathcal{G}^{(\nabla,f,a)}$ in Proposition 4.2, in which ∇ is replaced by D. Hence, from the relation (34), we have

$$
\begin{aligned}
(\mathscr{D}_{X^c}Y^c)_{(x,\omega)} &= (D_XY)^c_{(x,\omega)} + C_\omega\big((DX)(DY) + (DY)(DX)\big)_{(x,\omega)} \\
&\quad + C_\omega\big(R_x(., X)Y + R_x(., Y)X\big)_{(x,\omega)} \\
&\quad - \frac{1}{a}\big\{df(Y)X^c + df(X)Y^c\big\}_{(x,\omega)} \\
&\quad + \big\{\mu(X,Y) + \mu(Y,X)\big\}_{(x,\omega)},
\end{aligned} \tag{43}
$$

for arbitrary $X, Y \in \chi(M)$, where

$$
\begin{aligned}
\mu(X,Y) &= -\frac{1}{a}\big\{2df(Y)C_\omega(DX) + \omega(Y)\big(i_{df}(D_X\cdot) - i_{df}(DX)\big)^v \\
&\quad + \omega(D_XY)(df)^v + df(D_XY)L\big\} \tag{44} \\
&\quad + \frac{2}{a^2}\big\{df(X)df(Y)L + df(X)\omega(Y)(df)^v\big\}.
\end{aligned}
$$

From Remark 2.1 (i), it follows that in relation (43), the right hand side splits into the complete and the vertical part as

$$
(\mathscr{D}_{X^c}Y^c)_{(x,\omega)} = (D_XY)^c_{(x,\omega)} - \frac{1}{a}\big\{df(Y)X^c + df(X)Y^c\big\}_{(x,\omega)} \tag{45}
$$
$$
+ \text{(the vertical part)},
$$

for each $X, Y \in \chi(M)$. From (29), it follows that any integral curve of X is a classical geodesic (resp. an F-geodesic) with respect to D if and only if $D_X X = 0$ (resp. $D_X X = FX$). From (45), since $Xf = 0$, we obtain

$$(\mathscr{D}_{X^c} X^c)_{(x,\omega)} = (D_X X)^c_{(x,\omega)} + \text{(the vertical part)}, \qquad (46)$$

for each $X \in \chi(M)$. In the hypothesis, we made an assumption on the left hand side of (46), namely

$$(\mathscr{D}_{X^c} X^c)_{(x,\omega)} = 0 \quad \left(\text{resp. } (\mathscr{D}_{X^c} X^c)_{(x,\omega)} = F^c X^c_{(x,\omega)}\right), \qquad (47)$$

at each point $(x, \omega) \in T^*M$. If we replace (47) in (46) and based on Remark 2.1 (ii) and (9), we obtain

$$(D_X X)^c_{(x,\omega)} = 0 \quad \left(\text{resp. } (D_X X)^c_{(x,\omega)} = (FX)^c_{(x,\omega)}\right), \qquad (48)$$

at each point $(x, \omega) \in T^*M$. From Remark 2.1 (iii), it follows that the relation (48) is equivalent to

$$(D_X X)_x = 0 \quad \left(\text{resp. } (D_X X)_x = (FX)_x\right), \qquad (49)$$

at each point $x \in M$, which complete the proof. $\qquad\qquad\qquad\qquad\square$

Acknowledgment

The authors are very indebted to both the anonymous referee and the editor for careful reading of the manuscript and useful suggestions.

References

[1] T. Adachi, Kähler magnetic fields on a Kähler manifold of negative curvature, *Diff. Geom. Appl.*, **29**, S2–S8 (2011).

[2] C. L. Bejan and S. L. Druta-Romaniuc, F-geodesics on Manifolds, *Filomat*, **29**, no.10, 2367–2379 (2015).

[3] C. L. Bejan and O. Kowalski, On some differential operators on natural Riemann extensions, *Ann. Glob. Anal. Geom.*, **48**, 171–180 (2015).

[4] M. Brozos-Vazquez and E. Garcia-Rio, Four-dimensional neutral signature self-dual gradient Ricci solitons, *Indiana University Mathematics Journal*, **65**, no.6, 1921–1943 (2016).

[5] T. Higa, Weyl Manifolds and Einstein-Weyl Manifolds, *Comment. Math. Univ. St. Pauli*, **42**, no.2, 143–160 (1993).

[6] I. Kolar and P. W. Michor and J. Slovak, *Natural operations in differential geometry*, Springer, New York, 1993.

[7] O. Kowalski and M. Sekizawa, On natural Riemann extensions, *Publ. Math. Debrecen*, **78**, no.3-4, 709–721 (2011).

[8] O. Kowalski and M. Sekizawa, Almost Osserman structures on natural Riemann extensions, *Diff. Geom. Appl.*, **31**, no.1, 140–149 (2013).

[9] E. M. Patterson and A. G. Walker, Riemannian extensions, *Quart. J. Math. Oxford Ser.*, **2**, no.3, 19–28 (1952).

[10] M. Sekizawa, Natural transformations of affine connections on manifolds to metrics on cotangent bundles. *In Proceedings of 14th Winter School on Abstract Analysis (Srni, 1986) Rend. Circ. Mat. Palermo*, **14**, 129–142 (1987).

[11] T. J. Willmore, *An introduction to differential geometry*, Clarendon Press, Oxford 1959.

[12] K. Yano and E. M. Paterson, Vertical and complete lifts from a manifold to its cotangent bundle, *J. Math. Soc. Japan*, **19**, 91–113 (1967).

Received December 23, 2016
Revised   March 10, 2017

Contemporary Perspectives
in Differential Geometry
and its Related Fields   67 – 78

THE GEOMETRY OF ORBITS
OF HERMANN TYPE ACTIONS

Osamu IKAWA

Department of Mathematics and Physical Sciences,
Faculty of Arts and Sciences, Kyoto Institute of Technology,
Matsugasaki, Sakyoku, Kyoto 606-8585, Japan
E-mail: ikawa@kit.ac.jp

The notion of Hermann type actions is an analogy of that of Hermann actions. Hermann type actions are defined in the pseudo-Riemmanian category while Hermann actions are defined in the Riemannan category. It is known that Hermann actions have nice properties such as hyperpolarity. We study some nice properties of Hermann type actions like hyperpolar actions in the Riemannian category.

Keywords: Hermann type action; symmetric triad.

1. Introduction

This is an announcement of a part of results obtained in not-yet-published papers [1] and [2]. This research is a joint work with Kurando Baba (Tokyo university of science) and Atsumu Sasaki (Tokai University).

First we give the definition of Hermann type actions. Let G be a non-compact connected semisimple Lie group with finite center, and σ be an involutive automorphism of G. From now, we say involution instead of involutive automorphism for short. We denote by $F(\sigma, G)$ the fixed point subgroup of G by σ and set $H = F(\sigma, G)_0$, where $F(\sigma, G)_0$ is the identity component of $F(\sigma, G)$. It is known by Cartan that there exists a Cartan involution θ of G such that θ and σ commute with each other ([10, Theorem. 6.16]). Set $K = F(\theta, G)$. Then K is a maximal compact subgroup of G since the center of G is finite ([10, Theorem. 6.31]), and the coset manifold G/K has a structure of a symmetric space of noncompact type. The Lie group $G \cong G \times G/\Delta(G)$ and the coset manifold G/H have structures of pseudo-Riemannian symmetric spaces. In this paper we study the natural isometric actions as follows:

(1) The K-action on the pseudo-Riemannian symmetric space G/H,
(2) The $K \times H$-action on G,

(3) The H-action on the Riemannian symmetric space G/K of noncompact type.

In his paper [9], N. Koike called these the *Hermann type actions* after Robert Hermann. In this paper we study some nice properties of Hermann type actions like hyperpolar actions in the Riemannian category.

The author would like thank the referee for reading carefully the manuscript and pointing out some mistakes.

2. The geometry of orbits of Hermann type actions

We retain the notation as in Section 1 and denote by $\mathfrak{g}$ the Lie algebra of G. The involutions σ and θ on G induces involutions on $\mathfrak{g}$. We denote them by the same symbols. We define a subspace $\mathfrak{g}^{-\sigma,-\theta}$ by

$$\mathfrak{g}^{-\sigma,-\theta} = \{X \in \mathfrak{g} \mid -\sigma(X) = -\theta(X) = X\}. \tag{1}$$

Take a maximal abelian subspace $\mathfrak{a}$ of $\mathfrak{g}^{-\sigma,-\theta}$. If we set $A = \exp \mathfrak{a}$ then A is a connected abelian subgroup of G. Rossmann and Flensted-Jensen independently proved the following very important theorem:

Theorem 2.1 (Flensted-Jensen [7] and Rossmann [13]).
$G = KAH = HAK$.

Proof. This proof is due to F.-Jensen[7]. Denote by $\mathfrak{g} = \mathfrak{h} \oplus \mathfrak{q}$ the eigenspace decomposition by σ, and by $\mathfrak{g} = \mathfrak{k} \oplus \mathfrak{p}$ the Cartan decomposition by θ. Since $\mathfrak{p} \cap \mathfrak{q} = \mathfrak{g}^{-\sigma,-\theta}$ is a Lie triple system of $\mathfrak{p}$, we have, by Mostow's theorem ([5, Chap. VI, Theorem 1.4]),

$$G = K \exp(\mathfrak{p} \cap \mathfrak{q}) \exp(\mathfrak{p} \cap \mathfrak{h}). \tag{2}$$

Define a closed subgroup $G^{\theta\sigma}$ of G by

$$G^{\theta\sigma} = F(\theta\sigma, G) = \{g \in G \mid \theta\sigma(g) = g\}.$$

Then θ induces a Cartan involution of $G^{\theta\sigma}$, and the identity component $F(\theta, G^{\theta\sigma})_0$ of the fixed point subgroup $F(\theta, G^{\theta\sigma})$ is equal to $(H \cap K)_0$. A well-known result of symmetric spaces ([5, Lemma 6.3]) implies

$$\mathfrak{p} \cap \mathfrak{q} = \mathrm{Ad}(K \cap H)\mathfrak{a}. \tag{3}$$

By (2) and (3) we have

$$G = K(K \cap H) \exp \mathfrak{a}(K \cap H) \exp(\mathfrak{p} \cap \mathfrak{h}) = KAH.$$

Hence $G = G^{-1} = (KAH)^{-1} = HAK$. ☐

Theorem 2.1 means that A can be considered as a canonical form of the Hermann type actions. Denote by π_H the natural projection from G onto G/H, and by $K\backslash G/H$ the space of K-orbits of G/H. By Theorem 2.1 we can identify the orbit space $K\backslash G/H$ with $\pi_H(A)$ over an equivalence relation:

$$K\backslash G/H = \pi_H(A)/\sim,$$

where $x \sim y$ $(x, y \in \pi_H(A))$ means x and y are transformed each other by the K-action. Since exponential mapping exp from $\mathfrak{a}$ to A is surjective, we can also identify $K\backslash G/H$ with $\mathfrak{a}/\sim$. Here $X \sim Y$ $(X, Y \in \mathfrak{a})$ means $\exp X \sim \exp Y$ in the above sense. If we suitably select a representative of each element in $\mathfrak{a}$, we can identify $\mathfrak{a}/\sim$ with a certain connected closed convex subset in $\mathfrak{a}$. Recently we developed a theory of generalized duality and a theory of symmetric triad with multiplicities. Using these theories we can explicitly describe the orbit space $K\backslash G/H = \mathfrak{a}/\sim$. Further, with respect to each orbit, we get the following:

Theorem 2.2 ([1, 2]). *For the K-action on the pseudo-Riemannian symmetric space G/H the following statements hold.*

(1) *Any K-orbit is a pseudo-Riemannian submanifold in G/H.*
(2) *The induced symmetric bilinear form on the normal space $T_{gH}^{\perp} K(gH)$ of the K-orbit is positive definite for all g in G.*
(3) $\pi_H(A)$ *is a flat totally geodesic submanifold in G/H.*
(4) $\pi_H(A)$ *meets every K-orbit orthogonally with respect to the pseudo-Riemannian metric on G/H.*

We give a sketch of the proof. The third assertion follows from Theorem 4.3 and Proposition 4.4 in [8, Chap. XI]. Theorem 2.1 implies that $\pi_H(A)$ meets every K-orbit. The triple (G, H, K) defines a symmetric triad of $\mathfrak{a}$ (see [6] the definition of symmetric triads). We can express the tangent space $T_{gH}K(gH)$ and normal space $T_{gH}^{\perp}K(gH)$ using the symmetric triad. Using the expression we can prove the rest.

Theorem 2.3 ([1, 2]). *For the $K \times H$-action on G the following statements hold.*

(1) *Any $K \times H$-orbit is a pseudo-Riemannian submanifold in G.*
(2) *The induced symmetric bilinear form on the normal space of every $K \times H$-orbit is positive definite.*
(3) A *meets every $K \times H$-orbit orthogonally with respect to the Killing form on G.*

We give a sketch of the proof. Theorem 2.1 implies that A meets every $K \times H$-orbit. We can prove the rest using the symmetric triad of $\mathfrak{a}$ which is determined by (G, H, K).

Similarly to Theorems 2.2 and 2.3, concerning to H-action on G/K we get the following:

Theorem 2.4 ([1, 2]). *For the H-action on the symmetric space G/K of noncompact type the following statements hold.*

(1) $\pi_K(A)$ *is a flat totally geodesic submanifold in G/K, where we denote by π_K the natural projection from G onto G/K.*
(2) $\pi_K(A)$ *meets every H-orbit orthogonally.*

We emphasize that the generalized duality and symmetric triads with multiplicities are used in the proofs of the rest. Roughly speaking, in the Riemannian category the isometric action which satisfies the theorems above is called a *hyperpolar action*. There are many studies on hyperpolar actions, while there are few studies on Hermann type actions in the pseudo-Riemannian category. The author believes that it is important to study on this direction. To explain the usefulness of the above theorems, we apply it to concrete examples. It is the purpose of this paper.

Example 2.1. Denote by $\mathbb{C}^{p+q}$ the complex Euclidean space of complex dimension $p+q$, and by $\{e_1, \ldots, e_p, e_{p+1}, \ldots, e_{p+q}\}$ its canonical basis. Take an ordered basis of $\mathbb{R}^{2p+2q}$ as follows:

$$\{e_1, \ldots, e_p, \sqrt{-1}e_1, \ldots, \sqrt{-1}e_p, e_{p+1}, \ldots, e_{p+q}, \sqrt{-1}e_{p+1}, \ldots, \sqrt{-1}e_{p+q}\}.$$

The complex structure on $\mathbb{C}^{p+q}$ naturally induces the complex structure J on $\mathbb{R}^{2p+2q}$, which is represented by the matrix

$$J = \left(\begin{array}{cc|cc} 0 & -1_p & & \\ 1_p & 0 & & \\ \hline & & 0 & -1_q \\ & & 1_q & 0 \end{array} \right)$$

with respect to the above ordered basis of $\mathbb{R}^{2p+2q}$. We define a nondegenerate symmetric bilinear form $\langle\ ,\ \rangle$ on $\mathbb{R}^{2p+2q}$ as follows: For

$$x = \sum_{i=1}^{p}(x_i e_i + x_i'\sqrt{-1}e_i) + \sum_{j=1}^{q}(x_{p+j}e_{p+j} + x_{p+j}'\sqrt{-1}e_{p+j}),$$

$$y = \sum_{i=1}^{p}(y_i e_i + y_i'\sqrt{-1}e_i) + \sum_{j=1}^{q}(y_{p+j}e_{p+j} + y_{p+j}'\sqrt{-1}e_{p+j}),$$

we set

$$\langle x, y \rangle = \sum_{i=1}^{p}(x_i y_i + x_i' y_i') - \sum_{j=1}^{q}(x_{p+j} y_{p+j} + x_{p+j}' y_{p+j}').$$

From now we say the scalar product instead of nondegenerate symmetric bilinear form for simplicity. Set

$$I_{2p,2q} = \begin{pmatrix} 1_{2p} & \\ & -1_{2q} \end{pmatrix},$$

where we denote by 1_{2p} the identity matrix of degree $2p$. If we regard x and y as column vectors, then we can write

$$\langle x, y \rangle = {}^t x \, I_{2p,2q} \, y.$$

The scalar product $\langle \ , \ \rangle$ on $\mathbb{R}^{2p+2q}$ and the complex structure J satisfy the relation

$$\langle Jx, Jy \rangle = \langle x, y \rangle.$$

We simply denote the pair $(\mathbb{R}^{2p+2q}, \langle \ , \ \rangle)$ by $\mathbb{R}^{2p+2q}_{2p}$ and define a closed subgroup $O(2p, 2q)$ of $GL(2p + 2q, \mathbb{R})$ by

$$O(2p, 2q) = \left\{ g \in GL(2p + 2q, \mathbb{R}) \mid \langle gx, gy \rangle = \langle x, y \rangle \text{ for any } x, y \in \mathbb{R}^{2p+2q} \right\}$$
$$= \left\{ g \in GL(2p + 2q, \mathbb{R}) \mid {}^t g I_{2p,2q} g = I_{2p,2q} \right\}.$$

We see that the determinant of each element in $O(2p, 2q)$ is equal to 1 or -1 by the second equality above. Based on the fact, we define $SO(2p, 2q)$ by

$$SO(2p, 2q) = SL(2p + 2q, \mathbb{R}) \cap O(2p, 2q).$$

We denote by $G = SO(2p, 2q)_0$ the connected component of $SO(2p, 2q)$. Then

$$G = SO(2p, 2q)_0$$
$$= \left\{ g = \begin{pmatrix} g_{11} & g_{12} \\ g_{21} & g_{22} \end{pmatrix} \in SO(2p, 2q) \; \middle| \; \begin{array}{l} g_{11} \in GL(2p, \mathbb{R}), \\ g_{22} \in GL(2q, \mathbb{R}), \\ |g_{11}| > 0, \ |g_{22}| > 0, \\ g_{12}, {}^t g_{21} \in M(2p, 2q, \mathbb{R}) \end{array} \right\},$$

which is a connected noncompact simple Lie group. Define an involution θ on G by

$$\theta(g) = I_{2p,2q} \, g \, I_{2p,2q}^{-1}.$$

Then $K := F(\theta, G) = SO(2p) \times SO(2q)$. Since K is a maximal compact subgroup of G, we find that θ is a Cartan involution. Thus G/K is a Riemannian symmetric space of noncompact type. We explain that G/K can be seen as a set of subspaces in $\mathbb{R}_{2p}^{2p+2q}$. Denote by $G_{2p}(\mathbb{R}_{2p}^{2p+2q})$ the set of oriented positive definite real vector space of dimension $2p$ in $\mathbb{R}^{2p+2q}$. The dimension $2p$ means the maximal dimension such that the scalar product is positive definite on it. The action of G on $\mathbb{R}^{2p+2q}$ naturally induces an action of G on $G_{2p}(\mathbb{R}_{2p}^{2p+2q})$. The action of G on $G_{2p}(\mathbb{R}_{2p}^{2p+2q})$ is transitive. The isotropy subgroup of G at $\langle e_1, \dots, e_p, \sqrt{-1}e_1, \dots, \sqrt{-1}e_p \rangle_{\mathbb{R}} \in G_{2p}(\mathbb{R}_{2p}^{2p+2q})$ is equal to K. Thus G/K can be identified with $G_{2p}(\mathbb{R}_{2p}^{2p+2q})$ as

$$ G/K \ni gK \longleftrightarrow g\langle e_1, \dots, e_p, \sqrt{-1}e_1, \dots, \sqrt{-1}e_p \rangle_{\mathbb{R}} \in G_{2p}(\mathbb{R}_{2p}^{2p+2q}). $$

Set $\sigma(g) = JgJ^{-1}$ for $g \in G$. Then $\sigma(g)$ is in G. Thus σ is an involution of G. Set $H = F(\sigma, G) = \{g \in G \mid \sigma(g) = g\}$. Then we have

$$ H = \left\{ \begin{pmatrix} g_{11} & -g_{21} & g_{13} & -g_{23} \\ g_{21} & g_{11} & g_{23} & g_{13} \\ g_{31} & -g_{41} & g_{33} & -g_{43} \\ g_{41} & g_{31} & g_{43} & g_{33} \end{pmatrix} \middle| \begin{array}{l} g_{11}, g_{21} \in M(p, \mathbb{R}), \ g_{33}, g_{43} \in M(q, \mathbb{R}), \\ g_{13}, g_{23} \in M(p, q, \mathbb{R}), \ g_{31}, g_{41} \in M(q, p, \mathbb{R}), \\ \begin{pmatrix} g_{11} + \sqrt{-1}g_{21} & g_{13} + \sqrt{-1}g_{23} \\ g_{31} + \sqrt{-1}g_{41} & g_{33} + \sqrt{-1}g_{43} \end{pmatrix} \in U(p, q) \end{array} \right\} $$

$$ \cong U(p, q) = \{g \in GL(p+q, \mathbb{C}) \mid {}^t g I_{p,q} \bar{g} = I_{p,q}\}, $$

where $I_{p,q} = \begin{pmatrix} 1_p & \\ & -1_q \end{pmatrix}$. It is easy to see that σ and θ commute with each other. We consider the orbit space $U(p, q) \backslash G_{2p}(\mathbb{R}_{2p}^{2p+2q})$, that is, the space of all orbits of $U(p, q)$ on $G_{2p}(\mathbb{R}_{2p}^{2p+2q})$. The orbit space $U(p, q) \backslash G_{2p}(\mathbb{R}_{2p}^{2p+2q})$ can be expressed as the double coset space as follows:

$$ U(p, q) \backslash G_{2p}(\mathbb{R}_{2p}^{2p+2q}) = H \backslash G / K. $$

Without loss of generality we may assume that $p \leq q$. In order to describe the orbit space explicitly we prepare the following: For $x_1, x_2, \dots, x_p \in \mathbb{R}$ we define $V(x_1, \dots, x_p) \in G_{2p}(\mathbb{R}_{2p}^{2p+2q})$ by

$$ V(x_1, \dots, x_p) = \{u_1, \dots, u_p, v_1, \dots, v_p\}_{\mathbb{R}} $$

where we put

$$ u_i = (\cosh x_i)e_i + (\sinh x_i)e_{p+i}, \quad v_i = (\cosh x_i)\sqrt{-1}e_i - (\sinh x_i)\sqrt{-1}e_{p+i}. $$

For example when $x_1 = x_2 = \dots = x_p = 0$ then

$$ V(0, \dots, 0) = \{e_1, \dots, e_p, \sqrt{-1}e_1, \dots, \sqrt{-1}e_p\}_{\mathbb{R}}, $$

which is a complex subspace of $(\mathbb{R}_{2p}^{2p+2q}, J) = \mathbb{C}^{p+q}$.

In this setting we give an alternative proof of the following J. Mealy's result in 1989.

Proposition 2.1 (Mealy (1989) [11, 12]).

$$U(p,q)\backslash G_{2p}(\mathbb{R}^{2p+2q}_{2p}) = \{V(x_1,\ldots,x_p) \mid x_1 \geq x_2 \geq \cdots \geq x_p \geq 0\}.$$

Proof. The Lie algebra $\mathfrak{g} = \mathfrak{so}(2p,2q)$ of $G = SO(2p,2q)_0$ is expressed as

$$\mathfrak{g} = \left\{ \begin{pmatrix} X_{11} & X_{12} \\ {}^t X_{12} & X_{22} \end{pmatrix} \, \middle| \, \begin{array}{l} X_{11} \in \mathfrak{so}(2p), \ X_{22} \in \mathfrak{so}(2q), \\ X_{12} \in M(2p,2q,\mathbb{R}) \end{array} \right\}.$$

The subspace $\mathfrak{g}^{-\sigma,-\theta}$ defined by (1) is given by

$$\mathfrak{g}^{-\sigma,-\theta} = \left\{ \begin{pmatrix} O_{2p} & Y \\ {}^t Y & O_{2q} \end{pmatrix} \, \middle| \, Y = \begin{pmatrix} y_{11} & y_{12} \\ y_{12} & -y_{11} \end{pmatrix}, \text{where } y_{11}, y_{12} \in M(p,q,\mathbb{R}) \right\}.$$

Hence, we set $Y_i = E_{i,2p+i} - E_{p+i,2p+q+i} + E_{2p+i,i} - E_{2p+q+i,p+i}$ ($i = 1, 2, \ldots, p$) and define $\mathfrak{a} = \mathbb{R}Y_1 + \cdots + \mathbb{R}Y_p$. Then $\mathfrak{a}$ is a maximal abelian subspace of $\mathfrak{g}^{-\sigma,-\theta}$. For $X = \sum_{i=1}^{p} x_i Y_i \in \mathfrak{a}$ a simple calculation implies that

$$\exp X = \sum_{i=1}^{p} \{(\cosh x_i)Y_i^2 + (\sinh x_i)Y_i\}.$$

Thus we get

$$\exp X \ (V(0,\ldots,0)) = V(x_1,\ldots,x_p).$$

By Theorem 2.1 we have

$$U(p,q)\backslash G_{2p}(\mathbb{R}^{2p+2q}_{2p}) = \{V(x_1,\ldots,x_p)\}/\sim.$$

Denote by $\mathfrak{S}_p$ the symmetric group of degree p. Then $\mathfrak{S}_p$ can be regarded as a subgroup of $H \cong U(p,q)$ as follows. Let σ be in $\mathfrak{S}_p$. For $1 \leq i \leq p$ and $1 \leq j \leq q$ set $\sigma(e_i) = e_{\sigma(i)}$ and $\sigma(e_{p+j}) = e_{p+j}$. If we extend σ to a complex linear transformation on $\mathbb{C}^{p+q}$ then σ is in $U(p,q)$. Then we have $V(x_{\sigma(1)},\ldots,x_{\sigma(p)}) = \sigma(V(x_1,\ldots,x_p))$. Hence we get the assertion. $\square$

Our proof is based on Rossmann's theorem, while Mealy's proof was based on linear algebra. He applied the result to calibrated geometry in the pseudo-Riemannian category. We will explain it.

We define a 2-form Ω on $(\mathbb{R}^{2p+2q}_{2p}, J)$ by

$$\Omega(x,y) = \langle x, Jy \rangle.$$

Then the exterior p-th power $\Omega^p = \Omega \wedge \cdots \wedge \Omega$ of Ω is a $2p$-form on $(\mathbb{R}^{2p+2q}_{2p}, J)$, and $\left| \frac{1}{p!}\Omega^p(V) \right|$ is well-defined for V in $G_{2p}(\mathbb{R}^{2p+2q}_{2p})$. The following result due to Mealy immediately follows from Proposition 2.1.

Corollary 2.1 (Backwards Wirtinger Inequality (Mealy [11, 12])).
$\left|\frac{1}{p!}\Omega^p(V)\right| \geq 1$ for any $V \in G_{2p}(\mathbb{R}_{2p}^{2p+2q})$. The equality holds if and only if V is a complex subspace.

Proof. Since the action of $U(p, q)$ on $(\mathbb{R}_{2p}^{2p+2q}, J)$ commutes with J, and the scalar product $\langle \ , \ \rangle$ is invariant under the action, we have $|\Omega^p(V)| = |\Omega^p(gV)|$ for $V \in G_{2p}(\mathbb{R}_{2p}^{2p+2q})$ and $g \in U(p, q)$. Hence it is sufficient to prove the assertion when V is equal to $V(x_1, \cdots, x_p)$. A simple calculation implies the following:

$$\left|\frac{1}{p!}\Omega^p(V(x_1, \cdots, x_p))\right| = \prod_{j=1}^{p} \cosh(2x_j) \geq 1.$$

The equality holds if and only if $x_1 = \cdots = x_p = 0$, that is, V is a complex subspace. □

Mealy extended Harvey and Lawson's notion of a calibration on a Riemannian manifold to the pseudo-Riemannian category. Using the backward Wirtinger inequality he proved that the volume maximizing property of a maximally positive $2p$-dimensional submanifold of $(\mathbb{R}_{2p}^{2p+2q}, J)$ in a sense.

Proposition 2.1 immediately leads the following corollary.

Corollary 2.2. *There does not exist a real $2p$-dimensional vector space V of $\mathbb{R}_{2p}^{2p+2q}$ ($p \leq q$) which has the following property: The nondegenerate symmetric bilinear form $\langle \ , \ \rangle$ is positive definite on V, and V is totally real (that is, $\langle V, JV \rangle = \{0\}$).*

Proof. If it were to exist such a vector space V, then $\left|\frac{1}{p!}\Omega^p(V)\right| = 0$, which would be a contradiction to the backwards Wirtinger inequality. □

For a reference we state the usual Wirtinger inequality. Let the triple $(\mathbb{R}^{2n}, J, (\ , \ ))$ be the $2n$-dimensional Euclidean space with the canonical complex structure J and the canonical Hermitian inner product $(\ , \ )$. Denote by Ω the Kaehler form on $\mathbb{R}^{2n}$, which is defined by

$$\Omega(x, y) = (x, Jy) \quad \text{for} \quad x, y \in \mathbb{R}^{2n}.$$

Denote by $G_{2k}^{\mathbb{R}}(\mathbb{R}^{2n})$ the Grassmann manifold consisting of all real $2k$-dimensional vector subspace in $\mathbb{R}^{2n}$. For simplicity assume that $k \leq n$. Then $\left|\frac{1}{k!}\Omega^k(V)\right|$ is well-defined for $V \in G_{2k}^{\mathbb{R}}(\mathbb{R}^{2n})$. The usual Wirtinger inequality ([4, Theorem 7.12]) says

$$(0 \leq) \left|\frac{1}{k!}\Omega^k(V)\right| \leq 1.$$

Here $\left|\frac{1}{k!}\Omega^k(V)\right| = 1$ if and only if V is a complex subspace ($JV = V$). And $\left|\frac{1}{k!}\Omega^k(V)\right| = 0$ if and only if $V \perp JV$.

Using this Harvey and Lawson proved that each complex submanifold of $\mathbb{C}^n$ is volume minimizing ([4, Corollary 7.13]).

We will give another example.

Example 2.2. For $\mathbb{C}^n = \langle e_1, \ldots, e_n \rangle_{\mathbb{C}}$ set

$$\mathbb{R}^{2n} = \langle e_1, \ldots, e_n, \sqrt{-1}e_1, \ldots, \sqrt{-1}e_n \rangle_{\mathbb{R}}.$$

For $1 \leq j \leq n$ set $u_j = e_j$ and $u_{j+n} = \sqrt{-1}e_j$. Let M be the set of all (positive definite) inner products on $\mathbb{R}^{2n}$. It is clear that M can be identified with the set $\mathrm{Sym}^+(2n, \mathbb{R})$ of all positive definite symmetric matrices of degree $2n$. The action of $GL(2n, \mathbb{R})$ on $\mathbb{R}^{2n}$ naturally induces the action of $GL(2n, \mathbb{R})$ on M. That is,

$$(g\langle \, , \, \rangle)(x, y) = \langle g^{-1}x, g^{-1}y \rangle \quad \text{for} \quad g \in GL(2n, \mathbb{R}), \langle \, , \, \rangle \in M, x, y \in \mathbb{R}^{2n}.$$

The action of $GL(2n, \mathbb{R})$ on M is transitive. The isotropy subgroup of $GL(2n, \mathbb{R})$ on M at the canonical inner product is equal to $O(2n)$. Thus

$$M = GL(2n, \mathbb{R})/O(2n) \cong \mathrm{Sym}^+(2n, \mathbb{R}).$$

Since $GL(2n, \mathbb{R})$ is not semisimple we need a little idea. Define a subset M_0 of M by

$$M_0 = \{\langle \, , \, \rangle \in M \mid \det(\langle u_i, u_j \rangle) = 1\}.$$

Then we have

$$M_0 = SL(2n, \mathbb{R})/SO(2n) \cong \{X \in \mathrm{Sym}^+(2n, \mathbb{R}) \mid \det X = 1\}.$$

Here $G = SL(2n, \mathbb{R})$ is a noncompact connected simple Lie group. Define an involution θ on G by $\theta(g) = {}^t g^{-1}$. Then $K := F(\theta, G) = SO(2n)$. Since K is a maximal compact subgroup of G, we find that θ is a Cartan involution. Thus M_0 has a structure of a Riemannian symmetric space of noncompact type. We define an involution σ on G by

$$\sigma(g) = J_n^{-1} g J_n,$$

where the matrix $J_n = \begin{pmatrix} & -1_n \\ 1_n & \end{pmatrix}$ represents the complex structure on $\mathbb{R}^{2n}$. Then θ and σ commute with each other, and the fixed point subgroup $F(\sigma, G)$ of G is given by

$$H := F(\sigma, G) = \left\{ g = \begin{pmatrix} g_{11} & -g_{12} \\ g_{12} & g_{11} \end{pmatrix} \in GL(2n, \mathbb{R}) \,\middle|\, |g| = 1 \right\}.$$

The structure of H as a Lie group is described by the following lemma (but it is not used later).

Lemma 2.1. *The mapping* $F : U(1) \times SL(n, \mathbb{C}) \to H$ *defined by*

$$U(1) \times SL(n, \mathbb{C}) \ni (e^{\sqrt{-1}t}, h) \mapsto \begin{pmatrix} \mathrm{Re}(e^{\sqrt{-1}t}h) & -\mathrm{Im}(e^{\sqrt{-1}t}h) \\ \mathrm{Im}(e^{\sqrt{-1}t}h) & \mathrm{Re}(e^{\sqrt{-1}t}h) \end{pmatrix} \in H$$

is a Lie homomorphism from $U(1) \times SL(n, \mathbb{C})$ *onto* H. *The kernel* $\mathrm{Ker}F$ *is isomorphic to the center* $Z(SL(n, \mathbb{C}))$ *of* $SL(n, \mathbb{C})$.

Proof. In the proof we denote by $\det(A)$ the determinant of $A \in M(n, \mathbb{C})$, and by $|z|$ the absolute value of $z \in \mathbb{C}$ to avoid confusion. First note that

$$\det \begin{pmatrix} A & -B \\ B & A \end{pmatrix} = \left| \det(A + \sqrt{-1}B) \right|^2 \quad \text{for} \quad A, B \in M(n, \mathbb{R}).$$

Thus $F(e^{\sqrt{-1}t}, h)$ is in H for $(e^{\sqrt{-1}t}, h) \in U(1) \times SL(n, \mathbb{C})$. It is clear that F is a homomorphism. We show that F is surjective. In fact, since

$$1 = \det(g) = \left| \det(g_{11} + \sqrt{-1}g_{12}) \right|^2 \quad \text{for} \quad g = \begin{pmatrix} g_{11} & -g_{12} \\ g_{12} & g_{11} \end{pmatrix} \in H,$$

there exist $t \in \mathbb{R}$ such that $\det(g_{11} + \sqrt{-1}g_{12}) = e^{\sqrt{-1}nt}$. Then $e^{-\sqrt{-1}t}(g_{11} + \sqrt{-1}g_{12})$ is in $SL(n, \mathbb{C})$ and $F(e^{\sqrt{-1}t}, e^{-\sqrt{-1}t}(g_{11} + \sqrt{-1}g_{12})) = g$. Thus F is surjective. The kernel is given by

$$\mathrm{Ker}F = \left\{ (e^{\sqrt{-1}t}, h) \in U(1) \times SL(n, \mathbb{C}) \mid e^{\sqrt{-1}t}h = 1_n \right\}$$

$$= \left\{ (e^{\sqrt{-1}t}, h) \in U(1) \times SL(n, \mathbb{C}) \mid h = e^{-\sqrt{-1}t}1_n, \ t \in \frac{2\pi}{n}\mathbb{Z} \right\}$$

$$= \left\{ \left(z^{-1}, \mathrm{diag}(z, \cdots, z) \right) \mid z^n = 1 \right\} \cong Z(SL(n, \mathbb{C})).$$

We hence get the conclusion.   □

Proposition 2.2. *For any inner product* $\langle \, , \, \rangle$ *on* $\mathbb{R}^{2n} = \mathbb{C}^n$ *which satisfies* $\det(\langle u_i, u_j \rangle) = 1$, *there exist* $g \in H$ *and* $a_1 \geq a_2 \geq \cdots \geq a_n \geq 0$ *such that*

$$\langle x, y \rangle' := (g\langle \, , \, \rangle)(x, y) = \sum_{i=1}^{n} (e^{-2a_i}x_iy_i + e^{2a_i}x_{n+i}y_{n+i}). \tag{4}$$

Further $\langle \, , \, \rangle'$ *is Hermite if and only if* $a_1 = \cdots = a_n = 0$.

Proof. The subspace $\mathfrak{g}^{-\sigma,-\theta}$ defined by (1) is given by

$$\mathfrak{g}^{-\sigma,-\theta} = \left\{ \begin{pmatrix} X & Y \\ Y & -X \end{pmatrix} \ \middle| \ \begin{matrix} X, Y \in M(n, \mathbb{R}), \\ {}^t X = X, {}^t Y = Y \end{matrix} \right\}.$$

Hence, we set $Y_i = E_{ii} - E_{n+i,n+i}$ $(i = 1, 2, \ldots, n)$ and define $\mathfrak{a} = \mathbb{R}Y_1 + \cdots + \mathbb{R}Y_n$. Then $\mathfrak{a}$ is a maximal abelian subspace of $\mathfrak{g}^{-\sigma,-\theta}$. For $X = \sum_{i=1}^{n} a_i Y_i \in \mathfrak{a}$ a simple calculation implies that

$$\exp X = \begin{pmatrix} e^{a_1} & & & & & \\ & \ddots & & & & \\ & & e^{a_n} & & & \\ \hline & & & e^{-a_1} & & \\ & & & & \ddots & \\ & & & & & e^{-a_n} \end{pmatrix}.$$

Denote by $\langle\ ,\ \rangle_0$ the canonical inner product on $\mathbb{R}^{2n}$. Then for $x, y \in \mathbb{R}^{2n}$ we have

$$\langle\ ,\ \rangle_{(a_1,\ldots,a_n)}(x, y) := ((\exp X)\langle\ ,\ \rangle)(x, y) = \langle(\exp X)^{-1}x, (\exp X)^{-1}y\rangle$$

$$= \sum_{i=1}^{n}(e^{-2a_i}x_i y_i + e^{2a_i}x_{n+i}y_{n+i}).$$

For any permutation $\sigma \in \mathfrak{S}_n$ the two inner products $\langle\ ,\ \rangle_{(a_{\sigma(1)},\cdots,a_{\sigma(n)})}$ and $\langle\ ,\ \rangle_{(a_1,\cdots,a_n)}$ can be transformed each other by the action of H. Thus we get (4). The last assertion is clear. $\qquad\square$

If we identify the set of inner products with that of matrices, then we get the following:

Corollary 2.3. *For any positive definite symmetric matrix X of degree $2n$ which satisfies the condition* $\det X = 1$, *there exist* $g \in H$ *and* $a_1 \geq a_2 \geq \cdots \geq a_n \geq 0$ *such that*

$${}^t g^{-1} X g^{-1} = \begin{pmatrix} e^{-2a_1} & & & & & \\ & \ddots & & & & \\ & & e^{-2a_n} & & & \\ \hline & & & e^{2a_1} & & \\ & & & & \ddots & \\ & & & & & e^{2a_n} \end{pmatrix}.$$

In the usual linear algebra we learned canonical forms under the action $X \mapsto gXg^{-1}$. Since the eigenpolynomials are invariant under the action above, the set of eigenvalues of X is invariant under the action. But the eigenpolynomials are not invariant under the action $X \mapsto {}^t g^{-1} X g^{-1}$ in the corollary above. So the set of eigenvalues of X are not invariant. By this reason the author thinks that the result in the corollary is not trivial at all. In fact he cannot prove the corollary in the category of linear algebra. He feels that the theory of Hermann type actions is deeper than the theory of linear algebra.

References

[1] K. Baba, O. Ikawa and A. Sasaki, A duality between symmetric pairs and compact symmetric triads, in preparation.

[2] K. Baba, O. Ikawa and A. Sasaki, An alternative proof for Berger's classification of semisimple pseudo-Riemannian symmetric pairs from the viewpoint of compact symmetric triads, in preparation.

[3] R. Harvey and H. B. Lawson, Jr., Calibrated geometries, *Acta Math.*, **148**, 47–157 (1982).

[4] R. Harvey, *Spinors and calibrations*, Academic press, 1990.

[5] S. Helgason, *Differential geometry, Lie groups, and symmetric spaces*, Academic Press, 1978.

[6] O. Ikawa, The geometry of symmetric triad and orbit spaces of Hermann actions, *J. Math. Soc. Japan*, **63**, 79–136 (2011).

[7] M. Flensted-Jensen, Spherical functions on a real semisimple Lie group, A method of reduction to the complex case, *J. Fuct. Anal.*, **39**, 106–146. (1978)

[8] S. Kobayashi and K. Nomizu, *Foundations of differential geometry vol. II*, Interscience publishers, 1969.

[9] N. Koike, Actions of Hermann type and proper complex equifocal submanifolds, *Osaka J. Math.*, **42** 599–611 (2005).

[10] A. W. Knapp, *Lie groups beyond an introduction second edition*, Birkhäuser, 2005.

[11] J. Mealy, Calibrations on semi-Riemannian manifolds, Ph. D. Thesis, Rice Univ., Houston, 1989.

[12] J. Mealy, Volume maximization in semi-Riemannian manifolds, *Indiana Univ. Math. J.*, **40**, 793–814 (1991).

[13] W. Rossmann, The structure of semisimple symmetric spaces, *Canad. J. Math.*, **31**, 157–180 (1979).

Received November 22, 2016
Revised   January 31, 2017

FUNDAMENTAL RELATIONSHIP BETWEEN
CARTAN IMBEDDINGS OF TYPE A
AND HOPF FIBRATIONS

Hideya HASHIMOTO

Department of Mathematics, Meijo University,
Nagoya 468-8502, Japan
E-mail: hhashi@meijo-u.ac.jp

Misa OHASHI

Department of Mathematics, Nagoya Institute of Technology,
Nagoya 466-8555, Japan
E-mail: ohashi.misa@nitech.ac.jp

The homogeneous space $SU(2)/SO(2)$ can be considered as the Grassmaniann manifold of all totally real 2-planes of the complex plane $\mathbb{C}^2$. Also the homogeneous space $SU(2)/S(U(1) \times U(1))$ can be considered as the Grassmaniann manifold of all complex lines of $\mathbb{C}^2$. There does not exist a canonical isomorphism from a totally real 2-plane to a complex line, though, they are isomorphic to a 2-dimensional real vector space. Nevertheless, we can give the isomorphism from $SU(2)/SO(2)$ to $SU(2)/S(U(1) \times U(1))$. The purpose of this paper is to give this isomorphism and the bundle map from $SU(2)$ to $SU(2)$ corresponding to this isomorphism related to the Hopf fibrations, explicitly. By using this isomorphism we give some applications of non-flat totally geodesic immersions from S^2 to symmetric spaces of Type A.

Keywords: Grassmaniann manifolds; Hopf fibrations; bundle isomorphisms.

1. Introduction

The homogeneous space $SU(2)/SO(2)$ can be considered as the Grassmaniann manifold of all totally real 2-planes of the complex plane $\mathbb{C}^2$. Also the homogeneous space $SU(2)/S(U(1) \times U(1))$ can be considered as the Grassmaniann manifold of all complex lines of $\mathbb{C}^2$. It is well-known that these two homogeneous spaces are isomorphic to a 2-dimensional sphere. However, there does not exist a canonical isomorphism from a totally real 2-plane to a complex line, though, they are isomorphic to a 2-dimensional real vector space. Nevertheless, we can give the isomorphism from $SU(2)/SO(2)$ to $SU(2)/S(U(1) \times U(1))$. This isomorphism can be considered as the Cartan imbeddings from symmetric spaces of Type A to $SU(2)$. The correspondence of this isomorphism is related to the Hopf fibrations. The one of the purpose of this paper is to write down this isomorphism explicitly as a bundle map from $SU(2)$ to $SU(2)$. By using this isomorphism we give some applications of non-flat totally geodesic immersions from S^2 to symmetric spaces of Type A.

2. Symmetric spaces of Type A and Cartan imbeddings

In this section, we give the Cartan involutions of Type A, and the decomposition of the Lie algebra $\mathfrak{su}(n)$ of $SU(n)$ by these involutions. Let $SU(n)$ be the special unitary group of degree n defined by

$$SU(n) = \left\{ g \in M_{n \times n}(\mathbb{C}) \mid {}^t\bar{g}g = I_n, \ \det g = 1 \right\}.$$

We denote by σ the Cartan involution (automorphisms) of $SU(n)$ which satisfies

$$\sigma^2 = id_{SU(n)} \qquad (\sigma \neq id_{SU(n)}).$$

Then the isotropy subgroup of σ is given by

$$K = \left\{ h \in SU(n) \mid \sigma(h) = h \right\}.$$

Since σ is an involution, the differential $\sigma_*|_e$ at the identity element e

$$\sigma_*|_e : \mathfrak{su}(n) \longrightarrow \mathfrak{su}(n)$$

has two eigenvalues ± 1. We get the Cartan decomposition of the Lie algebra $\mathfrak{su}(n)$ by σ

$$\mathfrak{su}(n) = \mathfrak{k} \oplus \mathfrak{p}$$

where

$$\mathfrak{k} = \left\{ X \in \mathfrak{su}(n) \mid \sigma_*|_e(X) = X \right\},$$
$$\mathfrak{p} = \left\{ X \in \mathfrak{su}(n) \mid \sigma_*|_e(X) = -X \right\}.$$

Then the subspace $\mathfrak{p}$ can be identified with the tangent space $T_{eK}(SU(n)/K)$ at the origin $eK \in SU(n)/K$. We define the inner product $\langle \, , \, \rangle$ on $\mathfrak{p}$

$$\langle X, Y \rangle = \frac{1}{4} \mathrm{Re}\,(\mathrm{tr}\, {}^t\bar{X}Y)$$

for $X, Y \in \mathfrak{p}$, and we extend this inner product whole on $SU(n)/K$ by left translation. Then this inner product defines a biinvariant metric on $SU(n)/K$. We write the same symbol $\langle \, , \, \rangle$ this inner product. The triple $\left(SU(n)/K, \ \sigma, \ \langle \, , \, \rangle \right)$ is called Riemmanian symmetric space of Type A.

There exist 3-type irreducible symmetric spaces of Type A, which are called of Types AI, AII and AIII. We recall the Cartan involutions of Type A and the Cartan decomposition of $\mathfrak{su}(n)$ by each involution. To represent these, we put

$$J = \begin{pmatrix} O_{n\times n} & -I_n \\ I_n & O_{n\times n} \end{pmatrix}, \quad I_{p,q} = \begin{pmatrix} I_p & O_{p\times q} \\ O_{q\times p} & -I_q \end{pmatrix}.$$

Then, the Cartan involutions and decompositions are given by

Type	Cartan involution	$\mathfrak{k}$	$\mathfrak{p}$
AI	$\sigma_{I,n}(g) = \bar{g}$ (outer)	$\mathfrak{so}(n)$	$\sqrt{-1}\, U$
AII	$\sigma_{II,2n}(g) = J\bar{g}J^{-1}$ (outer)	$\mathfrak{sp}(n)$	$\begin{pmatrix} Z_1 & Z_2 \\ \bar{Z}_2 & -\bar{Z}_1 \end{pmatrix}$
AIII	$\sigma_{III,(p,q)}(g) = I_{p,q}gI_{p,q}$ (inner)	$\mathfrak{s}(\mathfrak{u}(p) \times \mathfrak{u}(q))$	$\begin{pmatrix} O_{p\times p} & -{}^t\bar{Z} \\ Z & O_{q\times q} \end{pmatrix}$

where $U \in M_{n\times n}(\mathbb{R})$ satisfies ${}^tU = U$, $Z_1 \in \mathfrak{su}(n), Z_2 \in \mathfrak{so}(n,\mathbb{C})$ and $Z \in M_{q\times p}(\mathbb{C})$.

Next we define the Cartan imbedding as follows. Let σ be one of three Cartan involutions, and set the mapping $\tilde{\mathrm{Car}}_\sigma : SU(n) \to SU(n)$ for $g \in SU(n)$ as

$$\tilde{\mathrm{Car}}_\sigma(g) = g\,\sigma(g^{-1}).$$

Then $\tilde{\mathrm{Car}}_\sigma$ is an immersion and the isotropy subgroup at the identity e of the immesion $\tilde{\mathrm{Car}}_\sigma$ coincides with K. Therefore the map $\tilde{\mathrm{Car}}_\sigma$ induce an imbedding from $SU(n)/K$ to $SU(n)$, that is

$$\mathrm{Car}_\sigma : SU(n)/K \to SU(n).$$

We call this mapping Car_σ the Cartan imbedding. These Cartan imbeddings are given by

Type	G/K	G	$\mathrm{Car}_\sigma(gK)$
AI	$SU(n)/SO(n)$	$SU(2n)$	$g\,{}^tg$
AII	$SU(2n)/Sp(n)$	$SU(2n)$	$gJ\,{}^tg\,J^{-1}$
AIII	$SU(p+q)/S(U(p) \times U(q))$	$SU(p+q)$	$gI_{p,q}\,{}^t\bar{g}\,I_{p,q}$

3. Cartan imbeddings of $SU(2)$ and a Hopf fibration

We give the relationship between the 2-dimensional symmetric spaces of Types AI, AIII and the Hopf fibration $S^3 \to S^2$.

First we shall consider the 2-dimensional symmetric space of Type AIII: $SU(2)/S\bigl(U(1) \times U(1)\bigr)$, the mapping $\tilde{\mathrm{Car}}_{\sigma_{\mathrm{III}},(1,1)} : SU(2) \to SU(2)$ is given by

$$\tilde{\mathrm{Car}}_{\sigma_{\mathrm{III}},(1,1)}(g) = g\, I_{1,1}\, {}^t\bar{g}\, I_{1,1} = \begin{pmatrix} |a|^2 - |b|^2 & -2a\bar{b} \\ 2\bar{a}b & |a|^2 - |b|^2 \end{pmatrix}$$

for $g = \begin{pmatrix} a & -\bar{b} \\ b & \bar{a} \end{pmatrix} \in SU(2)$. Therefore the image of $\tilde{\mathrm{Car}}_{\sigma_{\mathrm{III}},(1,1)}$ is diffeomorphic to $\mathbb{CP}^1 \cong S^2$. In fact,

$$\tilde{\mathrm{Car}}_{\sigma_{\mathrm{III}},(1,1)}\bigl(SU(2)\bigr) = \left\{ \begin{pmatrix} z & -\bar{w} \\ w & \bar{z} \end{pmatrix} \in SU(2) \;\middle|\; \mathrm{Im}\, z = 0 \right\} \cong S^2 \cong \mathbb{CP}^1.$$

We set the basis of $\mathfrak{su}(2)$ as

$$E_1 = \begin{pmatrix} 0 & -1 \\ 1 & 0 \end{pmatrix}, \quad E_2 = \begin{pmatrix} i & 0 \\ 0 & -i \end{pmatrix}, \quad E_3 = \begin{pmatrix} 0 & i \\ i & 0 \end{pmatrix}.$$

Then we have

Lemma 3.1. *The map* $\tilde{\mathrm{Car}}_{\sigma_{\mathrm{III}},(1,1)}$ *satisfies the following;*

$$\tilde{\mathrm{Car}}_{\sigma_{\mathrm{III}},(1,1)}\bigl(g\exp(\theta E_2)\bigr) = \tilde{\mathrm{Car}}_{\sigma_{\mathrm{III}},(1,1)}\left(g \begin{pmatrix} e^{i\theta} & 0 \\ 0 & e^{-i\theta} \end{pmatrix}\right) = \tilde{\mathrm{Car}}_{\sigma_{\mathrm{III}},(1,1)}(g)$$

for $e^{i\theta} \in S^1$ *and* $g \in SU(2)$. *The differential of the map* $\tilde{\mathrm{Car}}_{\sigma_{\mathrm{III}},(1,1)}$ *at the origin e satisfies*

$$\bigl(\tilde{\mathrm{Car}}_{\sigma_{\mathrm{III}},(1,1)}\bigr)_* |e\, (E_1) = 2E_1,$$
$$\bigl(\tilde{\mathrm{Car}}_{\sigma_{\mathrm{III}},(1,1)}\bigr)_* |e\, (E_3) = -2E_3,$$
$$\bigl(\tilde{\mathrm{Car}}_{\sigma_{\mathrm{III}},(1,1)}\bigr)_* |e\, (E_2) = 0.$$

Proposition 3.1. *The map*

$$\tilde{\mathrm{Car}}_{\sigma_{\mathrm{III}},(1,1)} : SU(2) \to SU(2)$$

defines a Hopf fibration. For any (fixed) element $\begin{pmatrix} x & -\bar{w} \\ w & x \end{pmatrix}$ *of the image* $\tilde{\mathrm{Car}}_{\sigma_{\mathrm{III}},(1,1)}(SU(2))$, *the inverse image at this point, is diffeomorphic to S^1, and is given concretely as follows:*

(1) *If $x \neq -1$, then*

$$\tilde{\mathrm{Car}}_{\sigma_{\mathrm{III},(1,1)}}^{-1} \left(\begin{pmatrix} x & -\bar{w} \\ w & x \end{pmatrix} \right)$$

$$= \left\{ \frac{1}{\sqrt{2(1+x)}} \begin{pmatrix} x+1 & -\bar{w} \\ w & x+1 \end{pmatrix} \begin{pmatrix} e^{i\theta} & 0 \\ 0 & e^{-i\theta} \end{pmatrix} \in SU(2) \;\middle|\; e^{i\theta} \in S^1 \right\}$$

where $\begin{pmatrix} x \\ w \end{pmatrix} \in S^2 \subset \mathbb{R} \oplus \mathbb{C}$, *which satisfy* $x^2 + |w|^2 = 1$.

(2) *If $x \neq 1$, then*

$$\tilde{\mathrm{Car}}_{\sigma_{\mathrm{III},(1,1)}}^{-1} \left(\begin{pmatrix} x & -\bar{w} \\ w & x \end{pmatrix} \right)$$

$$= \left\{ \frac{1}{\sqrt{2(1-x)}} \begin{pmatrix} \bar{w} & -(1-x) \\ 1-x & w \end{pmatrix} \begin{pmatrix} e^{i\theta} & 0 \\ 0 & e^{-i\theta} \end{pmatrix} \in SU(2) \;\middle|\; e^{i\theta} \in S^1 \right\}$$

where $\begin{pmatrix} x \\ w \end{pmatrix} \in S^2 \subset \mathbb{R} \oplus \mathbb{C}$, *which satisfy* $x^2 + |w|^2 = 1$.

Proof. Let $g = \begin{pmatrix} a & -\bar{b} \\ b & a \end{pmatrix}$ be any element of $SU(2)$. We solve the following equation for given g:

$$\tilde{\mathrm{Car}}_{\sigma_{\mathrm{III},(1,1)}} \left(\begin{pmatrix} a & -\bar{b} \\ b & a \end{pmatrix} \right) = \begin{pmatrix} |a|^2 - |b|^2 & -2a\bar{b} \\ 2\bar{a}b & |a|^2 - |b|^2 \end{pmatrix} = \begin{pmatrix} x & -\bar{w} \\ w & x \end{pmatrix}.$$

Or equivalently,

$$|a|^2 - |b|^2 = x, \tag{1}$$

$$2\bar{a}b = w, \tag{2}$$

$$|a|^2 + |b|^2 = 1. \tag{3}$$

By (1) and (3), we get $2|a|^2 = 1 + x$. Therefore there exists $e^{i\theta} \in S^1$ such that

$$a = \sqrt{\frac{1+x}{2}} e^{i\theta}.$$

If $x \neq -1$, by (2), we see that

$$b = \frac{e^{i\theta} w}{\sqrt{2(1+x)}}.$$

In the same way, if $x \neq 1$, since $2|b|^2 = 1 - x$, then

$$a = \frac{e^{i\theta}\bar{w}}{\sqrt{2(1-x)}}, \quad b = \sqrt{\frac{1-x}{2}} e^{i\theta}.$$

We get the desired result. □

We describe the local sections on $\tilde{\mathrm{Car}}_{\sigma_{\mathrm{III},(1,1)}}(SU(2)) \simeq S^2$. We take the open subsets U_1, U_2 of $\tilde{\mathrm{Car}}_{\sigma_{\mathrm{III},(1,1)}}(SU(2))$ as follows

$$U_1 = \left\{ \begin{pmatrix} x & -\bar{w} \\ w & x \end{pmatrix} \in \tilde{\mathrm{Car}}_{\sigma_{\mathrm{III},(1,1)}}(SU(2)) \ \middle| \ x \neq -1 \right\},$$

$$U_2 = \left\{ \begin{pmatrix} x & -\bar{w} \\ w & x \end{pmatrix} \in \tilde{\mathrm{Car}}_{\sigma_{\mathrm{III},(1,1)}}(SU(2)) \ \middle| \ x \neq 1 \right\}.$$

Then we can easily see that $U_1 \cup U_2 = \tilde{\mathrm{Car}}_{\sigma_{\mathrm{III},(1,1)}}(SU(2)) \simeq SU(2)/S(U(1) \times U(1))$. By Proposition 3.1, we define local section on U_1

$$\Gamma_1^{\mathrm{III}} : U_1 \to SU(2)$$

such that

$$\Gamma_1^{\mathrm{III}}\left(\begin{pmatrix} x & -\bar{w} \\ w & x \end{pmatrix} \right) = \frac{1}{\sqrt{2(1+x)}} \begin{pmatrix} x+1 & -\bar{w} \\ w & x+1 \end{pmatrix} \tag{4}$$

for $\begin{pmatrix} x & -\bar{w} \\ w & x \end{pmatrix} \in U_1$. In the same way, we define local section on U_2

$$\Gamma_2^{\mathrm{III}} : U_2 \to SU(2)$$

such that

$$\Gamma_2^{\mathrm{III}}\left(\begin{pmatrix} x & -\bar{w} \\ w & x \end{pmatrix} \right) = \frac{1}{\sqrt{2(1-x)}} \begin{pmatrix} \bar{w} & x-1 \\ 1-x & w \end{pmatrix} \tag{5}$$

for $\begin{pmatrix} x & -\bar{w} \\ w & x \end{pmatrix} \in U_2$. By (4) and (5), we see that

$$\Gamma_1^{\mathrm{III}}\left(\begin{pmatrix} x & -\bar{w} \\ w & x \end{pmatrix} \right) \begin{pmatrix} e^{i\alpha} & 0 \\ 0 & e^{-i\alpha} \end{pmatrix} = \Gamma_2^{\mathrm{III}}\left(\begin{pmatrix} x & -\bar{w} \\ w & x \end{pmatrix} \right)$$

for $\begin{pmatrix} x & -\bar{w} \\ w & x \end{pmatrix} \in U_1 \cap U_2$, where $e^{i\alpha} = \dfrac{\bar{w}}{|w|}$ $(\alpha = \arg \bar{w})$. Note that $w \neq 0$ on $U_1 \cap U_2$. In fact, we have

$$\frac{1}{\sqrt{2(1+x)}} \begin{pmatrix} x+1 & -\bar{w} \\ w & x+1 \end{pmatrix} \begin{pmatrix} e^{i\alpha} & 0 \\ 0 & e^{-i\alpha} \end{pmatrix} = \frac{1}{\sqrt{2(1-x)}} \begin{pmatrix} \bar{w} & -(1-x) \\ 1-x & w \end{pmatrix}$$

for $\begin{pmatrix} x & -\bar{w} \\ w & x \end{pmatrix} \in U_1 \cap U_2$. Therefore, by Proposition 3.1 the Cartan imbedding of Type AⅢ

$$\tilde{\mathrm{Car}}_{\sigma_{\mathrm{III},(1,1)}} : SU(2) \to \tilde{\mathrm{Car}}_{\sigma_{\mathrm{III},(1,1)}}\big(SU(2)\big) \subset SU(2)$$

can be considered as a non-trivial S^1-fibre bundle over S^2 and this fibration coincides with the Hopf fibration. We note that $SU(2)/S(U(1) \times U(1)) \simeq \mathbb{CP}^1$ can be considered as a Grassmann manifold of all complex lines in $\mathbb{C}^2$.

Next we consider the relationship between the symmetric space $SU(2)/SO(2)$ of Type AI, the Cartan imbedding and the Hopf fibration. Let $\tilde{\mathrm{Car}}_{\sigma_{\mathrm{I},2}} : SU(2) \to SU(2)$ be the map defined by

$$\tilde{\mathrm{Car}}_{\sigma_{\mathrm{I},2}}(g) = g\,{}^t g = \begin{pmatrix} a^2 + \bar{b}^2 & ab - \bar{a}\bar{b} \\ ab - \bar{a}\bar{b} & \bar{a}^2 + b^2 \end{pmatrix} \tag{6}$$

for $g = \begin{pmatrix} a & -\bar{b} \\ b & \bar{a} \end{pmatrix} \in SU(2)$. Then we can easily see that the image $\tilde{\mathrm{Car}}_{\sigma_{\mathrm{III},(1,1)}}$ is diffeomorphic to

$$\tilde{\mathrm{Car}}_{\sigma_{\mathrm{I},2}}\big(SU(2)\big) = \left\{ \begin{pmatrix} z & -\bar{w} \\ w & \bar{z} \end{pmatrix} \in SU(2) \ \middle|\ \mathrm{Re}\, w = 0 \right\} \cong S^2.$$

Then we get

Lemma 3.2. *The map* $\tilde{\mathrm{Car}}_{\sigma_{\mathrm{I},2}}$ *satisfies the following:*

$$\tilde{\mathrm{Car}}_{\sigma_{\mathrm{I},2}}\big(g\exp(\theta E_1)\big) = \tilde{\mathrm{Car}}_{\sigma_{\mathrm{I},2}}\left(g\begin{pmatrix} \cos\theta & -\sin\theta \\ \sin\theta & \cos\theta \end{pmatrix}\right) = \tilde{\mathrm{Car}}_{\sigma_{\mathrm{I},2}}(g)$$

for $e^{i\theta} \in S^1$ *and* $g \in SU(2)$. *The differential of the map* $\tilde{\mathrm{Car}}_{\sigma_{\mathrm{I},2}}$ *at the origin* e *satisfies*

$$\big(\tilde{\mathrm{Car}}_{\sigma_{\mathrm{I},2}}\big)_*|_e(E_2) = 2E_2,$$
$$\big(\tilde{\mathrm{Car}}_{\sigma_{\mathrm{I},2}}\big)_*|_e(E_3) = 2E_3,$$
$$\big(\tilde{\mathrm{Car}}_{\sigma_{\mathrm{I},2}}\big)_*|_e(E_1) = 0.$$

We note that $SU(2)/SO(2) \simeq S^2$ can be considered as a Grassmann manifold of all totally real 2-planes in $\mathbb{C}^2$.

We define the bundle map from $SU(2)$ to $SU(2)$ related to the Cartan imbeddings $\tilde{\mathrm{Car}}_{\sigma_{\mathrm{I},2}}$ and $\tilde{\mathrm{Car}}_{\sigma_{\mathrm{III},(1,1)}}$. To do this we define $K_0 \in SU(2)$ by

$$K_0 = \frac{1}{\sqrt{2}}\begin{pmatrix} 1 & i \\ i & 1 \end{pmatrix}.$$

Then we define a map (an inner automorphism) $\Psi : SU(2) \to SU(2)$ by

$$\Psi(g) = \mathbf{Ad}\,(K_0)\,g := K_0\,g\,K_0^{-1} = \frac{1}{2}\begin{pmatrix} 1 & i \\ i & 1 \end{pmatrix} g \begin{pmatrix} 1 & -i \\ -i & 1 \end{pmatrix}$$

for $g \in SU(2)$.

Lemma 3.3. *The map Ψ satisfies followings:*

$$\Psi\big(\exp(\theta E_1)\big) = \Psi\left(\begin{pmatrix} \cos\theta & -\sin\theta \\ \sin\theta & \cos\theta \end{pmatrix}\right) = \begin{pmatrix} e^{i\theta} & 0 \\ 0 & e^{-i\theta} \end{pmatrix},$$

$$\Psi\big(\exp(\theta E_2)\big) = \Psi\left(\begin{pmatrix} e^{i\theta} & 0 \\ 0 & e^{-i\theta} \end{pmatrix}\right) = \begin{pmatrix} \cos\theta & \sin\theta \\ -\sin\theta & \cos\theta \end{pmatrix},$$

$$\Psi\big(\exp(\theta E_3)\big) = \Psi\left(\begin{pmatrix} \cos\theta & i\sin\theta \\ i\sin\theta & \cos\theta \end{pmatrix}\right) = \begin{pmatrix} \cos\theta & i\sin\theta \\ i\sin\theta & \cos\theta \end{pmatrix}.$$

Next we define a map

$$\psi = \Psi|_{\tilde{\mathrm{C}}\mathrm{ar}_{\sigma_{\mathrm{I},2}}(SU(2))} : \tilde{\mathrm{C}}\mathrm{ar}_{\sigma_{\mathrm{I},2}}\big(SU(2)\big) \to \tilde{\mathrm{C}}\mathrm{ar}_{\sigma_{\mathrm{III}},(1,1)}\big(SU(2)\big)$$

by (6), we can put

$$\psi(h) = \mathbf{Ad}\,(K_0)\,h$$

for $h \in \tilde{\mathrm{C}}\mathrm{ar}_{\sigma_{\mathrm{I},2}}\big(SU(2)\big)$. Then the map ψ induces a diffeomorphism from $\tilde{\mathrm{C}}\mathrm{ar}_{\sigma_{\mathrm{I},2}}\big(SU(2)\big) \cong S^2$ to $\tilde{\mathrm{C}}\mathrm{ar}_{\sigma_{\mathrm{III}},(1,1)}\big(SU(2)\big) \cong \mathbb{CP}^1$. In fact, we obtain

Proposition 3.2. *For any $g \in SU(2)$, the following holds:*

$$\tilde{\mathrm{C}}\mathrm{ar}_{\sigma_{\mathrm{III}},(1,1)} \circ \Psi(g) = \psi \circ \tilde{\mathrm{C}}\mathrm{ar}_{\sigma_{\mathrm{I},2}}(g). \tag{7}$$

Proof. The left hand side of (7) is

$$\tilde{\mathrm{C}}\mathrm{ar}_{\sigma_{\mathrm{III}},(1,1)} \circ \Psi(g)$$

$$= \Psi(g)\,I_{1,1}\,{}^t\overline{\Psi(g)}\,I_{1,1}$$

$$= \frac{1}{2}\begin{pmatrix} 1 & i \\ i & 1 \end{pmatrix} g \begin{pmatrix} 1 & -i \\ -i & 1 \end{pmatrix} I_{1,1} \,{}^t\overline{\left(\frac{1}{2}\begin{pmatrix} 1 & i \\ i & 1 \end{pmatrix} g \begin{pmatrix} 1 & -i \\ -i & 1 \end{pmatrix}\right)} I_{1,1}$$

$$= \frac{1}{2}\begin{pmatrix} 1 & i \\ i & 1 \end{pmatrix} g \begin{pmatrix} 1 & -i \\ -i & 1 \end{pmatrix} I_{1,1} \left(\frac{1}{2}\begin{pmatrix} 1 & i \\ i & 1 \end{pmatrix} {}^t\bar{g} \begin{pmatrix} 1 & -i \\ -i & 1 \end{pmatrix}\right) I_{1,1}$$

$$= \frac{1}{2}\begin{pmatrix} 1 & i \\ i & 1 \end{pmatrix} g \begin{pmatrix} 1 & i \\ -i & -1 \end{pmatrix} \left(\frac{1}{2}\begin{pmatrix} 1 & i \\ i & 1 \end{pmatrix} {}^t\bar{g} \begin{pmatrix} 1 & i \\ -i & -1 \end{pmatrix}\right)$$

$$= \frac{1}{2}\begin{pmatrix} 1 & i \\ i & 1 \end{pmatrix} g \begin{pmatrix} 0 & i \\ -i & 0 \end{pmatrix} {}^t\bar{g} \begin{pmatrix} 1 & i \\ -i & -1 \end{pmatrix}. \tag{8}$$

On the other hand, the right hand side of (7) is

$$\psi \circ \tilde{\mathrm{Car}}_{\sigma_{\mathrm{I},2}}(g) = \psi(g\,{}^{t}g) = \frac{1}{2}\begin{pmatrix} 1 & i \\ i & 1 \end{pmatrix} g\,{}^{t}g\begin{pmatrix} 1 & -i \\ -i & 1 \end{pmatrix}. \tag{9}$$

If we put $g = \begin{pmatrix} a & -\bar{b} \\ b & \bar{a} \end{pmatrix}$, then

$$\begin{pmatrix} 0 & i \\ -i & 0 \end{pmatrix}{}^{t}\bar{g}\begin{pmatrix} 1 & i \\ -i & -1 \end{pmatrix} = \begin{pmatrix} 0 & i \\ -i & 0 \end{pmatrix}\begin{pmatrix} \bar{a} & \bar{b} \\ -b & a \end{pmatrix}\begin{pmatrix} 1 & i \\ -i & -1 \end{pmatrix} = \begin{pmatrix} -bi & ai \\ -\bar{a}i & -\bar{b}i \end{pmatrix}\begin{pmatrix} 1 & i \\ -i & -1 \end{pmatrix}$$

$$= \begin{pmatrix} -bi + a & b - ai \\ -\bar{a}i - \bar{b} & \bar{a} + \bar{b}i \end{pmatrix} = \begin{pmatrix} a & b \\ -\bar{b} & \bar{a} \end{pmatrix}\begin{pmatrix} 1 & -i \\ -i & 1 \end{pmatrix}$$

$$= {}^{t}g\begin{pmatrix} 1 & -i \\ -i & 1 \end{pmatrix}. \tag{10}$$

By (8), (9) and (10), we obtain the desired result. □

By Lemma 3.3 and Proposition 3.2, we see that $\Psi : SU(2) \to SU(2)$ is a diffeomorphism and satisfies the following commutative diagram;

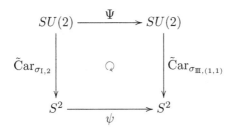

By using this diffeomorphism Ψ we construct the local section of the bundle $\tilde{\mathrm{Car}}_{\sigma_{\mathrm{I},2}} : SU(2) \to \tilde{\mathrm{Car}}_{\sigma_{\mathrm{I},2}}\big(SU(2)\big)$. To do this, we describe the S^1 fibre bundle structure related to $\tilde{\mathrm{Car}}_{\sigma_{\mathrm{I},2}}\big(SU(2)\big)$.

Proposition 3.3. *For any* $\begin{pmatrix} z & yi \\ yi & \bar{z} \end{pmatrix} \in \tilde{\mathrm{Car}}_{\sigma_{\mathrm{I},2}}(SU(2))$, *the inverse image at this point is given by the following:*

(1) *If* $\mathrm{Re}\,z \neq -1$, *then*

$$\tilde{\mathrm{Car}}_{\sigma_{\mathrm{I},2}}^{-1}\left(\begin{pmatrix} z & yi \\ yi & \bar{z} \end{pmatrix}\right)$$

$$= \left\{ \frac{1}{\sqrt{2(1 + \mathrm{Re}\,z)}}\begin{pmatrix} z + 1 & yi \\ yi & \bar{z} + 1 \end{pmatrix}\begin{pmatrix} \cos\theta & -\sin\theta \\ \sin\theta & \cos\theta \end{pmatrix} \,\Bigg|\, \begin{pmatrix} \cos\theta \\ \sin\theta \end{pmatrix} \in S^1 \right\}$$

where $\begin{pmatrix} z \\ y \end{pmatrix} \in S^2 \subset \mathbb{C} \oplus \mathbb{R}$, *which satisfy* $|z|^2 + y^2 = 1$.

(2) *If* $\operatorname{Re} z \neq 1$ *then*

$$\tilde{\mathrm{Car}}_{\sigma_{\mathrm{I},2}}^{-1} \left(\begin{pmatrix} z & yi \\ yi & \bar{z} \end{pmatrix} \right)$$

$$= \left\{ \frac{1}{\sqrt{2(1 - \operatorname{Re} z)}} \begin{pmatrix} i(z - 1) & y \\ -y & -i(\bar{z} - 1) \end{pmatrix} \begin{pmatrix} \cos\theta & -\sin\theta \\ \sin\theta & \cos\theta \end{pmatrix} \middle| \begin{pmatrix} \cos\theta \\ \sin\theta \end{pmatrix} \in S^1 \right\}$$

where $\begin{pmatrix} z \\ y \end{pmatrix} \in S^2 \subset \mathbb{C} \oplus \mathbb{R}$, *which satisfy* $|z|^2 + y^2 = 1$.

Proof. By Lemma 3.3, we have $\Psi^{-1} \left(\begin{pmatrix} e^{i\theta} & 0 \\ 0 & e^{-i\theta} \end{pmatrix} \right) = \begin{pmatrix} \cos\theta & -\sin\theta \\ \sin\theta & \cos\theta \end{pmatrix}$ for $e^{i\theta} \in S^1$. If we take $h \in \tilde{\mathrm{Car}}_{\sigma_{\mathrm{I},2}}(SU(2))$, then $\psi(h) \in U_j$ for some $j \in \{1, 2\}$. Therefore we obtain

$$\tilde{\mathrm{Car}}_{\sigma_{\mathrm{I},2}}^{-1}(h) = \left\{ \Psi^{-1} \left(\Gamma_j^{\mathrm{III}} \circ \psi(h) \begin{pmatrix} e^{i\theta} & 0 \\ 0 & e^{-i\theta} \end{pmatrix} \right) \middle| e^{i\theta} \in S^1 \right\}$$

$$= \left\{ \Psi^{-1} \left(\Gamma_j^{\mathrm{III}} \circ \psi(h) \right) \begin{pmatrix} \cos\theta & -\sin\theta \\ \sin\theta & \cos\theta \end{pmatrix} \middle| \begin{pmatrix} \cos\theta \\ \sin\theta \end{pmatrix} \in S^1 \right\}. \quad (11)$$

If $\psi(h) = \begin{pmatrix} x & -\bar{w} \\ w & x \end{pmatrix} \in U_1$, then

$$\Gamma_1^{\mathrm{III}} \left(\begin{pmatrix} x & -\bar{w} \\ w & x \end{pmatrix} \right) = \frac{1}{\sqrt{2(1 + x)}} \begin{pmatrix} x + 1 & -\bar{w} \\ w & x + 1 \end{pmatrix} \in \tilde{\mathrm{Car}}_{\sigma_{\mathrm{III},(1,1)}}^{-1} \left(\begin{pmatrix} x & -\bar{w} \\ w & x \end{pmatrix} \right).$$

On the other hand, the elment $\Psi^{-1} \circ \Gamma_1^{\mathrm{III}} \left(\begin{pmatrix} x & -\bar{w} \\ w & x \end{pmatrix} \right)$ is given by

$$\Psi^{-1} \circ \Gamma_1^{\mathrm{III}} \left(\begin{pmatrix} x & -\bar{w} \\ w & x \end{pmatrix} \right)$$

$$= \frac{1}{2} \begin{pmatrix} 1 & -i \\ -i & 1 \end{pmatrix} \left(\frac{1}{\sqrt{2(1 + x)}} \begin{pmatrix} x + 1 & -\bar{w} \\ w & x + 1 \end{pmatrix} \right) \begin{pmatrix} 1 & i \\ i & 1 \end{pmatrix}$$

$$= \frac{1}{\sqrt{2(1 + x)}} \begin{pmatrix} x + 1 - i(\operatorname{Re} w) & i(\operatorname{Im} w) \\ i(\operatorname{Im} w) & x + 1 + i(\operatorname{Re} w) \end{pmatrix}.$$

Then we have

$$\begin{pmatrix} x & -\bar{w} \\ w & x \end{pmatrix} = \psi \left(\begin{pmatrix} z & yi \\ yi & \bar{z} \end{pmatrix} \right) = \begin{pmatrix} \operatorname{Re} z & (\operatorname{Im} z) + yi \\ -(\operatorname{Im} z) + yi & \operatorname{Re} z \end{pmatrix}.$$

Since $x = \operatorname{Re} z \neq -1$, we get

$$\Psi^{-1}\left(\Gamma_1^{\mathrm{III}} \circ \psi\left(\begin{pmatrix} z & yi \\ yi & \bar{z} \end{pmatrix}\right)\right)$$

$$= \Psi^{-1}\left(\frac{1}{\sqrt{2(1+x)}}\begin{pmatrix} \operatorname{Re} z + 1 & -(\operatorname{Im} z) - yi \\ -(\operatorname{Im} z) + yi & \operatorname{Re} z + 1 \end{pmatrix}\right)$$

$$= \frac{1}{\sqrt{2(1+\operatorname{Re} z)}}\begin{pmatrix} z+1 & yi \\ yi & \bar{z}+1 \end{pmatrix}. \tag{12}$$

In the same way, if $\psi(h) = \begin{pmatrix} x & -\bar{w} \\ w & x \end{pmatrix} \in U_2$ then we get

$$\Gamma_2^{\mathrm{III}}\left(\begin{pmatrix} x & -\bar{w} \\ w & x \end{pmatrix}\right) = \frac{1}{\sqrt{2(1-x)}}\begin{pmatrix} \bar{w} & x-1 \\ 1-x & w \end{pmatrix} \in \check{\mathrm{Car}}_{\sigma_{\mathrm{III},(1,1)}}^{-1}\left(\begin{pmatrix} x & -\bar{w} \\ w & x \end{pmatrix}\right).$$

From this $\Psi^{-1} \circ \Gamma_2^{\mathrm{III}} \circ \psi(h)$ is given by

$$\Psi^{-1} \circ \Gamma_2^{\mathrm{III}}\left(\begin{pmatrix} x & -\bar{w} \\ w & x \end{pmatrix}\right)$$

$$= \frac{1}{\sqrt{2(1-x)}}\begin{pmatrix} (\operatorname{Re} w) - i(1-x) & \operatorname{Im} w \\ -(\operatorname{Im} w) & (\operatorname{Re} w) + i(1-x) \end{pmatrix}.$$

Therefore, if $\operatorname{Im} z \neq 1$ then we get

$$\Psi^{-1}\left(\Gamma_2^{\mathrm{III}} \circ \psi\left(\begin{pmatrix} z & yi \\ yi & \bar{z} \end{pmatrix}\right)\right)$$

$$= \Psi^{-1}\left(\frac{1}{\sqrt{2(1-\operatorname{Re} z)}}\begin{pmatrix} -(\operatorname{Im} z) - yi & (\operatorname{Re} z) - 1 \\ 1 - (\operatorname{Re} z) & -(\operatorname{Im} z) + yi \end{pmatrix}\right)$$

$$= \frac{1}{\sqrt{2(1-\operatorname{Re} z)}}\begin{pmatrix} i(z-1) & y \\ -y & -i(\bar{z}-1) \end{pmatrix}. \tag{13}$$

By (11), (12) and (13), we obtain the desired result. $\qquad\square$

Let V_1, V_2 be the open subsets of $\check{\mathrm{Car}}_{\sigma_{\mathrm{I},2}}(SU(2))$ defined by

$$V_1 = \left\{\begin{pmatrix} z & yi \\ yi & \bar{z} \end{pmatrix} \in \check{\mathrm{Car}}_{\sigma_{\mathrm{I},2}}(SU(2)) \;\middle|\; \operatorname{Re} z \neq -1\right\},$$

$$V_2 = \left\{\begin{pmatrix} z & yi \\ yi & \bar{z} \end{pmatrix} \in \check{\mathrm{Car}}_{\sigma_{\mathrm{I},2}}(SU(2)) \;\middle|\; \operatorname{Re} z \neq 1\right\}.$$

Then we can easily see that $V_1 \cup V_2 = \check{\mathrm{Car}}_{\sigma_{\mathrm{I},2}}(SU(2)) \simeq SU(2)/SO(2)$. The local sections are defined by:

(1) If $h = \begin{pmatrix} z & yi \\ yi & \bar{z} \end{pmatrix} \in V_1$, then

$$\Gamma_1^{\mathrm{I}}\left(\begin{pmatrix} z & yi \\ yi & \bar{z} \end{pmatrix}\right) = \frac{1}{\sqrt{2(1+\operatorname{Re}z)}}\begin{pmatrix} z+1 & yi \\ yi & \bar{z}+1 \end{pmatrix}.$$

(2) If $h \in V_2$, then

$$\Gamma_2^{\mathrm{I}}\left(\begin{pmatrix} z & yi \\ yi & \bar{z} \end{pmatrix}\right) = \frac{1}{\sqrt{2(1-\operatorname{Re}z)}}\begin{pmatrix} i(z-1) & y \\ -y & -i(\bar{z}-1) \end{pmatrix}.$$

We put $\alpha = \arg\left(-(\operatorname{Im}z + yi)\right)$ then the following holds

$$\Gamma_1^{\mathrm{I}}\left(\begin{pmatrix} z & yi \\ yi & \bar{z} \end{pmatrix}\right)\begin{pmatrix} \cos\alpha & -\sin\alpha \\ \sin\alpha & \cos\alpha \end{pmatrix} = \Gamma_2^{\mathrm{I}}\left(\begin{pmatrix} z & yi \\ yi & \bar{z} \end{pmatrix}\right)$$

for $\begin{pmatrix} z & yi \\ yi & \bar{z} \end{pmatrix} \in V_1 \cap V_2$.

4. Non-flat totally geodesic surfaces in symmetric spaces of Type A and Cartan imbeddings

We recall the fundamental results of irreducible representation of $SU(2)$. Let $V(d)$ be the complex vector space of all homogeneous polynomials of degree d of complex two variables $(z, w) \in \mathbb{C}^2$ defined by

$$V(d) = \operatorname{span}_{\mathbb{C}}\left\{z^{d-j}w^j \mid j \in \mathbb{Z},\ 0 \leq j \leq d\right\} \cong \mathbb{C}^{d+1}.$$

Let $\rho : SU(2) \to \operatorname{End}\{V(d)\}$ be the linear representation of $SU(2)$ defined by

$$\left(\rho(g)P\right)(z, w) = P\left({}^t\!\left(g^{-1}\begin{pmatrix} z \\ w \end{pmatrix}\right)\right) = P((z, w)\bar{g})$$

for $g \in SU(2)$. Then ρ implies a complex irreducible representation of $SU(2)$.

We define the homogeneous polynomial $P_k \in V(d)$ by

$$P_k(z, w) = \frac{z^{d-k}w^k}{\sqrt{k!(d-k)!}}$$

for $k \in \{0,\ 1,\ 2,\ \ldots,\ d\}$. Then

$$(P_0,\ P_1,\ \ldots,\ P_d) \tag{14}$$

is an orthonormal basis of $V(d)$ with repsect to the following Hermitian inner product $\langle \, , \, \rangle$. For any two homogeneous polynomials

$$f_1(z,w) = \sum_{k=0}^{d} a_k z^{d-k} w^k, \quad f_2(z,w) = \sum_{k=0}^{d} b_k z^{d-k} w^k \in V(d),$$

we put

$$\langle f_1, \, f_2 \rangle = \sum_{k=0}^{d} (d-k)! \, k! \, a_k \overline{b_k}.$$

If we put $n = d + 1$, then we define the representaion matrix $\mu_n(g)$ of $\rho(g)$ with respect to the orthonormal basis of (14) for $g \in SU(2)$ as

$$\begin{pmatrix} \rho(g)P_0 & \rho(g)P_1 & \cdots & \rho(g)P_d \end{pmatrix} = \begin{pmatrix} P_0 & P_1 & \cdots & P_d \end{pmatrix} \mu_n(g). \tag{15}$$

Then μ_n is a $SU(n)$-valued C^∞-function on $SU(2)$. In ([5]) K. Mashimo classified non-flat totally geodesic surfaces in symmetric spaces of classical type. The composition of Mashimo's immersion and the Cartan imbedding, we can describe the non-flat totally geodesic immersions (surfaces) in symmetric spaces of Type A by polynomials. In order to explain the construction of the composition of mappings, we prepare the followings:

Let p, q be the integers which satisfy $p = q$ or $p = q + 1$. Let $Q' \in O(p+q) \subset M_{(p+q) \times (p+q)}$ be the matrix defined by

$$(Q')_{ij} = \begin{cases} \delta_j^{2i-1} & (1 \le i \le p), \\ \delta_j^{2(i-p)} & (p+1 \le i \le p+q). \end{cases}$$

We put $\mu_{\mathbb{II},(p,q)}(g)$ as

$$\mu_{\mathbb{II},(p,q)}(g) = \mathbf{Ad}(Q')\mu_{p+q}(g).$$

Next we put $Q_n \in O(n)$ as

$$Q_n = \begin{pmatrix} 0 & \cdots & 0 & 1 \\ \vdots & & \reflectbox{$\ddots$} & 0 \\ 0 & \reflectbox{$\ddots$} & & \vdots \\ 1 & 0 & \cdots & 0 \end{pmatrix} \in O(n).$$

That is, $(Q_n)_{ij} = \delta_j^{n+1-i}$. We define the matrix of reducible representation $\mu_{\mathbb{II},2n}(g)$ of $SU(2)$ by

$$\mu_{\mathbb{II},2n}(g) = \mathbf{Ad}(I_n \oplus Q_n)\Delta\mu_n(g)$$

$$= \begin{pmatrix} I_n & O_{n \times n} \\ O_{n \times n} & Q_n \end{pmatrix} \begin{pmatrix} \mu_n(g) & O_{n \times n} \\ O_{n \times n} & \mu_n(g) \end{pmatrix} \begin{pmatrix} I_n & O_{n \times n} \\ O_{n \times n} & Q_n \end{pmatrix}^{-1}.$$

Then the Cartan imbeddings and representation matricies satisfy the following relations (in [3]):

(1) Let

$$\phi_{\mathrm{I}} : SU(2)/SO(2) \to SU(n)/SO(n)$$

be the totally geodesic immersion of $SU(n)/SO(n)$ give in [5] and

$$\mathrm{Car}_{\sigma_{\mathrm{I},n}} : SU(n)/SO(n) \to SU(n)$$

be the Cartan imbedding. The composition of two mappings

$$\varphi_{\mathrm{I}} = \mathrm{Car}_{\sigma_{\mathrm{I},n}} \circ \phi_{\mathrm{I}} : \tilde{\mathrm{Car}}_{\sigma_{\mathrm{I},2}}\big(SU(2)\big) \to SU(n)$$

satisfies

$$\varphi_{\mathrm{I}}(h) = \mu_n(h)$$

for $h \in \tilde{\mathrm{Car}}_{\sigma_{\mathrm{I},2}}\big(SU(2)\big)$.

(2) Let p, q be the integers which satisfy $p = q$ or $p = q + 1$. Let

$$\phi_{\mathrm{III}} : SU(2)/S(U(1) \times U(1)) \to SU(p+q)/S(U(p) \times U(q))$$

be the totally geodesic immersion of $SU(p+q)/S(U(p) \times U(q))$ and

$$\mathrm{Car}_{\sigma_{\mathrm{III},(p,q)}} : SU(p+q)/S(U(p) \times U(q)) \to SU(p+q)$$

be the Cartan imbedding. The composition of two mappings

$$\varphi_{\mathrm{III}} = \mathrm{Car}_{\sigma_{\mathrm{III},(p,q)}} \circ \phi_{\mathrm{III}} : \tilde{\mathrm{Car}}_{\sigma_{\mathrm{III},(1,1)}}\big(SU(2)\big) \to SU(p+q)$$

satisfies

$$\varphi_{\mathrm{III}}(k) = \mu_{\mathrm{III},(p,q)}(k)$$

for $k \in \tilde{\mathrm{Car}}_{\sigma_{\mathrm{III},(1,1)}}\big(SU(2)\big)$.

(3) The composition of totally geodesic immersion

$$\phi_{\mathrm{II}} : SU(2) \to SU(n)/Sp(n)$$

and the Cartan imbedding

$$\mathrm{Car}_{\sigma_{\mathrm{II},2n}} : SU(n)/Sp(n) \to SU(n)$$

is given by

$$\varphi_{\mathrm{II}} = \mathrm{Car}_{\sigma_{\mathrm{II},2n}} \circ \phi_{\mathrm{II}} : \tilde{\mathrm{Car}}_{\sigma_{\mathrm{III},(1,1)}}\big(SU(2)\big) \to SU(2n).$$

Then we have

$$\varphi_{\mathrm{II}}(k) = \mu_{\mathrm{II},2n}(k)$$

for $k \in \tilde{\mathrm{Car}}_{\sigma_{\mathrm{III},(1,1)}}\big(SU(2)\big)$.

We obtained the following theorem and its proof included in [3] with K. Suzuki and the authors.

Theorem 4.1. *The totally geodesic immersions* φ_{I}, φ_{II}, φ_{III} *satisfy the following relations.*

(1) *Take n for $p+q$. Let $F_{\mathrm{III}} : SU(n) \to SU(p+q)$ be the map defined by*

$$F_{\mathrm{III}}(g) = \mathbf{Ad}\left(Q' \mu_n(K_0)\right)(g)$$

for $g \in SU(n)$, where $K_0 = \dfrac{1}{\sqrt{2}} \begin{pmatrix} 1 & i \\ i & 1 \end{pmatrix}$. A diffeomorphism F_{III} satisfies

$$\varphi_{\mathrm{III}} = F_{\mathrm{III}} \circ \varphi_{\mathrm{I}} \circ \psi^{-1},$$

where

$$\psi : \mathrm{Car}_{\sigma_{\mathrm{I},2}}\big(SU(2)\big) \to \mathrm{Car}_{\sigma_{\mathrm{III},(1,1)}}\big(SU(2)\big)$$

and $\psi = \mathbf{Ad}(K_0^{-1})$.

(2) *Let $F_{\mathrm{II}} : SU(n) \to SU(2n)$ be the map defined by*

$$F_{\mathrm{II}}(g) = \mathbf{Ad}\big(I_n \oplus Q_n\big)\Delta\big(\mathbf{Ad}(\mu_n(K_0))\big)(g)$$

for $g \in SU(n)$. Then F_{II} is a homomorphism whcih satisfies

$$\varphi_{\mathrm{II}} = F_{\mathrm{II}} \circ \varphi_{\mathrm{I}} \circ \psi^{-1},$$

where

$$\psi : \mathrm{Car}_{\sigma_{\mathrm{I},2}}\big(SU(2)\big) \to \mathrm{Car}_{\sigma_{\mathrm{III},(1,1)}}\big(SU(2)\big),$$

and $\psi = \mathbf{Ad}(K_0^{-1})$.

References

[1] T. Fujimaru, A. Kubo and H, Tamaru, On totally geodesic surfaces in symmetric spaces of type AI, *Springer Proc. Math. Stat.*, **106**, 211–227, Springer, Tokyo, (2014).

[2] H. Hashimoto, and K. Suzuki, Hopf fibration and Cartan imbedding of type AI, in *Current Developments in Differential Geometry and its Related Fields*, T. Adachi, H. Hashimoto & M. Hristov eds., World Scientific, Singapole, 155–163, (2015).

[3] H. Hashimoto, M. Ohashi and K. Suzuki, On the Polynomial representations of non-flat totally geodesic surfaces in symmetric spaces of Type A, in preparation.

[4] S. Helgason, *Differential geometry, Lie group, and symmetric spaces*, AMS, (1978).

H. HASHIMOTO & M. OHASHI

[5] K. Mashimo, Non-flat totally geodesic surfaces of symmetric space of classical type, to appear.

[6] M. Takeuchi, *Lie groups 2,* Iwanami Shoten, Tokyo, (1984)

[7] J.A. Wolf, *Spaces of constant curvature.* Sixth edition. AMS Chelsa Publishing, Providence, RI, 2011.

Received March 23, 2017
Revised May 18, 2017

Contemporary Perspectives
in Differential Geometry
and its Related Fields   95 – 112

A STUDY ON TRAJECTORY-HORNS
FOR KÄHLER MAGNETIC FIELDS

In memory of my father Masayoshi Adachi.

Toshiaki ADACHI*

*Department of Mathematics, Nagoya Institute of Technology,
Nagoya 466-8555, Japan
E-mail: adachi@nitech.ac.jp*

A trajectory-horn associated with a geodesic is a variation of trajectories for a Kähler magnetic field where each of the trajectories joins the origin and another point of the geodesic. We study lengths of these trajectory-segments, inner products of the tangent vectors of the geodesic and trajectories, and angles of two trajectories at the origin of the geodesic. By comparing trajectory-horns with those on complex space forms, we estimate these quantities under an assumption that sectional curvatures are bounded from above.

Keywords: Kähler magnetic fields; trajectory-horns; tube-lengths; tube-cosines; embouchure angles; trajectory-harps; comparison theorems.

1. Introduction

On a complete Kähler manifold (M, J) with complex structure J, we take Kähler magnetic fields, which are closed 2-forms obtained as constant multiples of its Kähler form $\mathbb{B}_J$. A smooth curve γ parameterized by its arclength is said to be a *trajectory* for a Kähler magnetic field $\mathbb{B}_\kappa = \kappa \mathbb{B}_J$ ($\kappa \in \mathbb{R}$) if it satisfies the differential equation $\nabla_{\dot\gamma}\dot\gamma = \kappa J\dot\gamma$. Clearly, when $\kappa = 0$, it is a geodesic of unit speed. As J is parallel, it is a circle of geodesic curvature $|\kappa|$. Here, a smooth curve γ parameterized by its arclength is said to be a circle of geodesic curvature k (≥ 0) if it satisfies $\nabla_{\dot\gamma}\dot\gamma = kY$, $\nabla_{\dot\gamma}Y = -k\dot\gamma$ with a field Y of unit vectors along γ. In view of this Frenet-Serre's formula, we may say that circles are simplest curves next to geodesics. Since trajectories induce a dynamical system on the unit tangent bundle UM of M like geodesics ([1, 13]), we can say that trajectories are generalized objects which are closely related with the complex structure on the underlying Kähler manifold. The author therefore considers that they should

---

*The author is partially supported by Grant-in-Aid for Scientific Research (C)
(No. 16K05126), Japan Society for the Promotion of Science.

show some properties of the underlying manifold associated with its complex structure. The aim of this paper is to study trajectories by considering the relationship between geodesics and trajectories.

In this direction we studied trajectory-harps in [5–7] and [11]. A trajectory-harp is a smooth variation of geodesics associated with a trajectory (see §3 for precise definition). Roughly speaking, trajectory-harps show how trajectories diffuse from their initial points. Correspondingly, in this paper we study some basic properties of trajectory-horns. A trajectory-horn consists of a geodesic and trajectories each of which joins the initial point and another point of the geodesic, and forms a smooth variation of trajectories. Since magnetic Jacobi fields obtained by variation of trajectories are not orthogonal to trajectories ([2]), it is not easy to treat trajectory-horns. By support of comparison theorems on trajectory-harps given in [5, 6], we give estimates on tube-lengths and tube-cosines of trajectory-horns, which are lengths of these trajectory-segments joining two points of geodesics and inner products of the tangent vectors of geodesics and trajectories. Also we give estimates on their embouchure angles by applying a comparison theorem on magnetic Jacobi fields.

2. Trajectory-horns

In order to study the relationship between trajectories and geodesics we here introduce some kind of variations of trajectories. We here start by recalling some terminologies. Let M be a complete Kähler manifold. On this manifold every trajectory for a Kähler magnetic field is defined on the whole real line. We call a smooth curve a trajectory-segment if it is a restriction of a trajectory to a finite interval, and call it a trajectory half-line if it is a restriction of a trajectory to an infinite interval like $(-\infty, 0]$ and $[0, \infty)$. For the sake of simplicity, we say trajectory-segments and trajectory half-lines also as trajectories.

For a unit tangent vector $v \in U_pM$ we denote by γ_v^κ the trajectory for a Kähler magnetic field $\mathbb{B}_\kappa$ whose initial vector is v. The magnetic exponential map $\mathbb{B}_\kappa \exp_p : T_pM \to M$ of the tangent space at $p \in M$ is defined as

$$\mathbb{B}_\kappa \exp_p(w) = \begin{cases} \gamma_{w/\|w\|}^\kappa(\|w\|), & \text{when } w \neq 0_p, \\ p, & \text{when } w = 0_p. \end{cases}$$

Being different from the ordinary exponential map $\exp_p = \mathbb{B}_0 \exp_p$, this magnetic exponential map is not necessarily surjective. For example, on a complex Euclidean space $\mathbb{C}^n$, a trajectory for $\mathbb{B}_\kappa$ is a circle of radius $1/|\kappa|$ in the sense of Euclidean geometry, hence the image of $\mathbb{B}_\kappa \exp_p$ coincides with a closed ball of radius $1/|\kappa|$ centered at p. We can only say that when sectional curvatures of M satisfy $\mathrm{Riem}^M \le c < 0$ with some constant c and $|\kappa| \le \sqrt{|c|}$ the magnetic exponential map $\mathbb{B}_\kappa \exp_p$ is a covering map at an arbitrary point $p \in M$ ([5]). We set

$$\iota_p^\kappa = \sup\{\rho > 0 \mid \mathbb{B}_\kappa \exp_p|_{B_\rho(0_p)} \text{ is injective}\},$$

where $B_\rho(0_p)$ is the open ball of radius ρ centered at 0_p in $T_p M$, and call it the arc-radius of $\mathbb{B}_\kappa$-injectivity at p. Clearly $\iota_p^0 = \iota_p$ is the radius of injectivity at p. When M is simply connected and satisfies $\mathrm{Riem}^M \le c < 0$, if $|\kappa| \le \sqrt{|c|}$ then we have $\iota_p^\kappa = \infty$ for an arbitrary p.

A vector field Y along a trajectory γ for $\mathbb{B}_\kappa$ is said to be a *magnetic Jacobi field* for $\mathbb{B}_\kappa$ if it satisfies

$$\begin{cases} \nabla_{\dot\gamma} \nabla_{\dot\gamma} Y + \kappa J \nabla_{\dot\gamma} Y + R(Y, \dot\gamma)\dot\gamma = 0, \\ \langle \nabla_{\dot\gamma} Y, \dot\gamma \rangle = 0. \end{cases} \tag{2.1}$$

A magnetic Jacobi field is obtained from a variation of trajectories, and vice versa ([2]). Given a vector field X along γ we decompose it and denote as $X(t) = f_X(t)\dot\gamma(t) + g_X(t)J\dot\gamma(t) + X^\perp(t)$ with functions f_X, g_X and a vector field $X^\perp$ along γ which is orthogonal to both $\dot\gamma, J\dot\gamma$ at each point. We set $X^\sharp = g_X J\dot\gamma + X^\perp$. For a trajectory γ for $\mathbb{B}_\kappa$ we say t_c a magnetic conjugate value of $p = \gamma(0)$ along γ if there is a nontrivial magnetic Jacobi field Y satisfying $Y(0) = 0$ and $Y^\sharp(t_c) = 0$. We denote by $c_{p,\gamma}^\kappa$ the minimum positive magnetic conjugate value of p along γ if it exists. We set $c_{p,\gamma}^\kappa = \infty$ if we have no positive magnetic conjugate values of p along γ. We put $c_p^\kappa = \min\{c_{p,\gamma_v}^\kappa \mid v \in U_p M\}$ and call it the first $\mathbb{B}_\kappa$-magnetic conjugate value at p. We note that $c_p^0 = c_p$ denotes the first conjugate value at p.

Let $\sigma : [0, S] \to M$ be a geodesic segment or geodesic half-line satisfying $\sigma(s) \ne \sigma(0)$ for $0 < s < S$. We suppose that the image $\sigma([0, S])$ is contained in the image $\mathbb{B}_\kappa \exp_{\sigma(0)}(T_p M)$ of the magnetic exponential map for a non-trivial $\mathbb{B}_\kappa$ ($\kappa \ne 0$). A smooth variation $\beta_\sigma : [0, S] \times \mathbb{R} \to M$ of trajectories for $\mathbb{B}_\kappa$ is said to be a *trajectory-horn* for $\mathbb{B}_\kappa$ associated with σ if it satisfies the following conditions:

i) $\beta_\sigma(s, 0) = \sigma(0)$ for $0 \le s \le S$,

ii) when $s = 0$, the curve $t \mapsto \beta_\sigma(0, t)$ is the trajectory for $\mathbb{B}_\kappa$ of initial vector $\dot\sigma(0)$,

iii) when $0 < s < S$ the curve $t \mapsto \beta_\sigma(s, t)$ is a trajectory for $\mathbb{B}_\kappa$ joining $\sigma(0)$ and $\sigma(s)$.

When the image $\sigma([0, S])$ is contained in the trajectory-ball $\mathbb{B}_\kappa B_{\iota_p^\kappa}(p) = \mathbb{B}_\kappa \exp_p(B_{\iota_p^\kappa}(0_p))$ centered at $p = \sigma(0)$ and of arc-radius ι_p^κ, as we can join $\gamma(0)$ and $\gamma(t)$ by a unique minimizing trajectory for $\mathbb{B}_\kappa$, we have a unique trajectory-horn each of whose trajectory-segments has minimal length. Therefore, if $\sigma([0, S])$ lies in the trajectory-ball $\mathbb{B}_\kappa B_{c_p^\kappa}(p)$, we have a trajectory-horn by extending the above one.

For a trajectory-horn β_σ for $\mathbb{B}_\kappa$ associated with σ, we denote by $r_{\sigma,\kappa}(s)$ the arclength of the trajectory-segment $t \mapsto \beta_\sigma(s, t)$ joining $\sigma(0)$ and $\sigma(s)$ when $s > 0$, set $r_{\sigma,\kappa}(0) = 0$, and call it its *tube-length*. We set $\epsilon_{\sigma,\kappa}(s) = \langle \dot\sigma(s), \frac{\partial\beta_\sigma}{\partial t}(s, r_{\sigma,\kappa}(s)) \rangle$, $\mu_{\sigma,\kappa}(s) = \left\| \frac{\partial\beta_\sigma}{\partial s}(s, r_{\sigma,\kappa}(s)) \right\|$, and call them the *tube-cosine* and the *expansion* of β_σ, respectively. Trivially we have $\epsilon_{\sigma,\kappa}(0) = 1$ and $\mu_{\sigma,\kappa}(0) = 0$. The differential of the tube-length and the tube-cosine satisfy the following relations.

Lemma 2.1. *For a trajectory-horn β_σ for $\mathbb{B}_\kappa$ associated with a geodesic σ, we have*

$$\epsilon_{\sigma,\kappa}(s) \frac{dr_{\sigma,\kappa}}{ds}(s) + \left\langle \frac{\partial\beta_\sigma}{\partial s}(s, r_{\sigma,\kappa}(s)), \dot\sigma(s) \right\rangle = 1, \tag{2.2}$$

$$\frac{dr_{\sigma,\kappa}}{ds}(s) + \left\langle \frac{\partial\beta_\sigma}{\partial s}(s, r_{\sigma,\kappa}(s)), \frac{\partial\beta_\sigma}{\partial t}(s, r_{\sigma,\kappa}(s)) \right\rangle = \epsilon_{\sigma,\kappa}(s). \tag{2.3}$$

In particular, we have $\lim_{s\downarrow 0} \frac{dr_{\sigma,\kappa}}{ds}(s) = 1$.

Proof. As we have $\sigma(s) = \beta_\sigma\big(s, r_{\sigma,\kappa}(s)\big)$, we see

$$\dot\sigma(s) = \frac{\partial\beta_\sigma}{\partial s}(s, r_{\sigma,\kappa}(s)) + r'_{\sigma,\kappa}(s) \frac{\partial\beta_\sigma}{\partial t}(s, r_{\sigma,\kappa}(s)), \tag{2.4}$$

where $r'_{\sigma,\kappa} = dr_{\sigma,\kappa}/ds$. Taking the inner products of both sides of this equality with $\dot\sigma(s)$ and $\frac{\partial\beta_\sigma}{\partial t}(s, r_{\sigma,\kappa}(s))$ we get the equalities. $\square$

We denote by γ_s the trajectory $t \mapsto \beta_\sigma(s, t)$ for $\mathbb{B}_\kappa$, and put $Y_s(t) = \frac{\partial\beta_\sigma}{\partial s}(s, t)$, which is a magnetic Jacobi field for $\mathbb{B}_\kappa$ along γ_s. Since $\nabla_{\dot\gamma_s} Y_s$ is orthogonal to $\dot\gamma_s$, by direct computation and by (2.4) we have

Lemma 2.2. *For a trajectory-horn β_σ for $\mathbb{B}_\kappa$ associated with a geodesic σ, we have*

$$\frac{d\epsilon_{\sigma,\kappa}}{ds}(s) = \left\langle Y_s\big(r_{\sigma,\kappa}(s)\big), \big(\nabla_{\dot\gamma_s} Y_s\big)\big(r_{\sigma,\kappa}(s)\big) \right\rangle$$

$$+ \kappa \frac{dr_{\sigma,\kappa}}{ds}(s) \left\langle \dot\sigma(s), J\dot\gamma_s\big(r_{\sigma,\kappa}(s)\big) \right\rangle.$$

In particular, we have $\lim_{s\downarrow 0} \frac{d\epsilon_{\sigma,\kappa}}{ds}(s) = 0$.

Lemma 2.3. *For a trajectory-horn β_σ for $\mathbb{B}_\kappa$ associated with a geodesic σ, we have*

$$\lim_{s\downarrow 0}\frac{d^2 r_{\sigma,\kappa}}{ds^2}(s) = 0, \qquad \lim_{s\downarrow 0}\left(\nabla_{\dot\gamma_s}Y_s\right)(s) = -\frac{\kappa}{2}J\dot\gamma_0(0),$$

$$\lim_{s\downarrow 0}\frac{d^2 \epsilon_{\sigma,\kappa}}{ds^2}(s) = -\frac{\kappa^2}{4}, \qquad \lim_{s\downarrow 0}\frac{d\mu_{\sigma,\kappa}}{ds}(s) = \frac{|\kappa|}{2}.$$

Proof. Differentiating both sides of (2.3), as $\nabla_{\frac{\partial\beta_\sigma}{\partial t}}\frac{\partial\beta_\sigma}{\partial s}$ is orthogonal to $\frac{\partial\beta_\sigma}{\partial t}$, we have

$$r''_{\sigma,\kappa}(s) + \left\langle \left(\nabla_{\frac{\partial\beta_\sigma}{\partial s}}\frac{\partial\beta_\sigma}{\partial s}\right)(s, r_{\sigma,\kappa}(s)), \frac{\partial\beta_\sigma}{\partial t}(s, r_{\sigma,\kappa}(s))\right\rangle$$

$$+ \left\langle \frac{\partial\beta_\sigma}{\partial s}(s, r_{\sigma,\kappa}(s)), \left(\nabla_{\frac{\partial\beta_\sigma}{\partial t}}\frac{\partial\beta_\sigma}{\partial s}\right)(s, r_{\sigma,\kappa}(s))\right\rangle$$

$$+ \kappa r'_{\sigma,\kappa}(s)\left\langle \frac{\partial\beta_\sigma}{\partial s}(s, r_{\sigma,\kappa}(s)), J\frac{\partial\beta_\sigma}{\partial t}(s, r_{\sigma,\kappa}(s))\right\rangle = \epsilon'_{\sigma,\kappa}(s).$$

As $\frac{\partial\beta_\sigma}{\partial s}(s,0) = 0$, we obtain $\lim_{s\downarrow 0} r''_{\sigma,\kappa}(s) = 0$.

Differentiating both sides of (2.4) we have

$$0 = \nabla_{\dot\sigma}\dot\sigma(s)$$

$$= \left(\nabla_{\frac{\partial\beta_\sigma}{\partial s}}\frac{\partial\beta_\sigma}{\partial s}\right)(s, r_{\sigma,\kappa}(s)) + 2\,r'_{\sigma,\kappa}(s)\left(\nabla_{\frac{\partial\beta_\sigma}{\partial t}}\frac{\partial\beta_\sigma}{\partial s}\right)(s, r_{\sigma,\kappa}(s))$$

$$+ r''_{\sigma,\kappa}(s)\frac{\partial\beta_\sigma}{\partial t}(s, r_{\sigma,\kappa}(s)) + \kappa\, r'_{\sigma,\kappa}(s)\, J\frac{\partial\beta_\sigma}{\partial t}(s, r_{\sigma,\kappa}(s)),$$

hence obtain $\lim_{s\downarrow 0}\left(\nabla_{\dot\gamma_s}Y_s\right)(0) + (\kappa/2)\,J\dot\gamma_0(0) = 0$.

By Lemma 2.2 we have

$$\epsilon''_{\sigma,\kappa}(s) = \left\langle \left(\nabla_{\frac{\partial\beta_\sigma}{\partial s}}\frac{\partial\beta_\sigma}{\partial s}\right)(s, r_{\sigma,\kappa}(s)), \left(\nabla_{\frac{\partial\beta_\sigma}{\partial t}}\frac{\partial\beta_\sigma}{\partial s}\right)(s, r_{\sigma,\kappa}(s))\right\rangle$$

$$+ r'_{\sigma,\kappa}(s)\left\|\left(\nabla_{\frac{\partial\beta_\sigma}{\partial t}}\frac{\partial\beta_\sigma}{\partial s}\right)(s, r_{\sigma,\kappa}(s))\right\|^2$$

$$+ \left\langle \frac{\partial\beta_\sigma}{\partial s}(s, r_{\sigma,\kappa}(s)), \left(\nabla_{\frac{\partial\beta_\sigma}{\partial s}}\nabla_{\frac{\partial\beta_\sigma}{\partial t}}\frac{\partial\beta_\sigma}{\partial s}\right)(s, r_{\sigma,\kappa}(s))\right\rangle$$

$$+ r'_{\sigma,\kappa}(s)\left\langle \frac{\partial\beta_\sigma}{\partial s}(s, r_{\sigma,\kappa}(s)), \left(\nabla_{\frac{\partial\beta_\sigma}{\partial t}}\nabla_{\frac{\partial\beta_\sigma}{\partial t}}\frac{\partial\beta_\sigma}{\partial s}\right)(s, r_{\sigma,\kappa}(s))\right\rangle$$

$$+ \kappa\, r''_{\sigma,\kappa}(s)\left\langle \dot\sigma(s), J\frac{\partial\beta_\sigma}{\partial t}(s, r_{\sigma,\kappa}(s))\right\rangle$$

$$+ \kappa\, r'_{\sigma,\kappa}(s)\left\langle \dot\sigma(s), J\left(\nabla_{\frac{\partial\beta_\sigma}{\partial t}}\frac{\partial\beta_\sigma}{\partial s}\right)(s, r_{\sigma,\kappa}(s))\right\rangle$$

$$- \kappa^2\epsilon_{\sigma,\kappa}(s)\left(r'_{\sigma,\kappa}(s)\right)^2.$$

As $\lim_{s\downarrow 0}\big(\nabla_{\dot\gamma_s}Y_s\big)(s) = -(\kappa/2)\,J\dot\gamma_0(0)$, we get $\lim_{s\downarrow 0}\epsilon''_{\sigma,\kappa}(s) = -\kappa^2/4$.

By differentiating $\mu_{\sigma,\kappa}(s)^2 = \big\|\frac{\partial\beta_\sigma}{\partial s}\big(s, r_{\sigma,\kappa}(s)\big)\big\|^2$ twice, we obtain the last by using $\lim_{s\downarrow 0}\big(\nabla_{\dot\gamma_t}Y_s\big)(s) = -(\kappa/2)\,J\dot\gamma_0(0)$. □

3. Trajectory-horns on complex space forms

In this section we study trajectory-horns on a complex space form $\mathbb{C}M^n(c)$ of constant holomorphic sectional curvature c, which is a complex projective space $\mathbb{C}P^n(c)$, a complex Euclidean space $\mathbb{C}^n$ and a complex hyperbolic space $\mathbb{C}H^n(c)$ according as $c > 0$, $c = 0$ and $c < 0$. On $\mathbb{C}M^n(c)$, two trajectories γ_1, γ_2 for $\mathbb{B}_\kappa$ are congruent to each other in strong sense, that is, there is an isometry φ of $\mathbb{C}M^n(c)$ satisfying $\gamma_2(t) = \varphi \circ \gamma_1(t)$ for all t (see [1, 9]). Therefore tube-lengths, tube-cosines, tube-expansions do not depend on trajectory-horns for a given Kähler magnetic field on $\mathbb{C}M^n(c)$. We hence denote them as $r_\kappa(s; c)$, $\epsilon_\kappa(s; c)$ and $\mu_\kappa(s; c)$. Since a trajectory γ_1 for $\mathbb{B}_\kappa$ and a trajectory γ_2 for $\mathbb{B}_{-\kappa}$ are congruent to each other on $\mathbb{C}M^n(c)$, we see $r_\kappa(s; c) = r_{-\kappa}(s; c)$ and so on. Given a non-zero constant κ we set

$$d(\kappa; c) = \begin{cases} (2/\sqrt{c})\tan^{-1}\big(\sqrt{c}/|\kappa|\big), & \text{when } c > 0, \\ 2/|\kappa|, & \text{when } c = 0, \\ (2/\sqrt{|c|})\tanh^{-1}\big(\sqrt{|c|}/|\kappa|\big), & \text{when } c < 0 \text{ and } |\kappa| > \sqrt{|c|}, \\ \infty, & \text{when } c < 0 \text{ and } |\kappa| \le \sqrt{|c|}. \end{cases}$$

This satisfies $d(\kappa; c_1) > d(\kappa; c_2)$ when $c_1 < c_2$ and $d(\kappa_1; c) > d(\kappa_2; c)$ when $|\kappa_1| < |\kappa_2|$. Congruency of trajectories for a given $\mathbb{B}_\kappa$ on $\mathbb{C}M^n(c)$ guarantees that every trajectory ball $\mathbb{B}_\kappa B_\rho(p) = \mathbb{B}_\kappa \exp_p(B_\rho(0_p))$ of arc-radius ρ centered at p coincides with some geodesic ball centered at p. This number $d(\kappa; c)$ shows the radius of the geodesic ball whose closure coincides with $\mathbb{B}_\kappa \exp_p(T_pM)$.

Proposition 3.1. *On $\mathbb{C}^n$, for $0 \le s \le 2/|\kappa|$, we have the following:*

(1) *The tube-length is given by $r_\kappa(s; 0) = (2/|\kappa|)\sin^{-1}(|\kappa|s/2)$, hence is monotone increasing;*

(2) *The tube-cosine is given by $\epsilon_\kappa(s; 0) = \sqrt{1 - (\kappa^2 s^2/4)}$, hence is monotone decreasing;*

(3) *When $s < 2/|\kappa|$, the expansion is given by $\mu_\kappa(s; 0) = |\kappa|s/\sqrt{4 - \kappa^2 s^2}$, hence is monotone increasing and satisfies $\lim_{s\uparrow 2/|\kappa|}\mu_\kappa(s; 0) = \infty$.*

Proof. We are enough to study on $\mathbb{C}^1$. Since a trajectory γ for $\mathbb{B}_\kappa$ is a circle of radius $1/|\kappa|$ in the sense of Euclidean geometry, if a geodesic segment σ

joins two distinct points $\sigma(0) = \gamma(0)$ and $\sigma(s) = \gamma\big(r_\kappa(s; 0)\big)$ on γ, then it is a subtense for this circular arc. Considering the isosceles triangle of vertices $\sigma(0), \sigma(s)$ and the center p of γ, we see the central angle $\angle\big(\sigma(0)p\sigma(s)\big)$ is $2\sin^{-1}(|\kappa|s/2)$, hence the angle of circumference $\cos^{-1}\epsilon_\kappa(s; 0)$ is its half and the arclength $r_\kappa(s; 0)$ of the circular arc is $(2/|\kappa|)\sin^{-1}(|\kappa|s/2)$. We therefore have $\epsilon_\kappa(s; 0) = \cos\big(\sin^{-1}(|\kappa|s/2)\big) = \sqrt{1 - (\kappa^2 s^2/4)}$.

We take a geodesic $\sigma(s) = su$ of initial unit vector $u \in \mathbb{C} \cong T_o\mathbb{C}$ at the origin o. Since a trajectory γ for $\mathbb{B}_\kappa$ with $\gamma(0) = o$ and $\dot\gamma(0) = v$ is expressed as $\gamma(t) = (1/\kappa)\big\{\sin\kappa t - \sqrt{-1}(\cos\kappa t - 1)\big\}v$, we find by the expression of $\epsilon_\kappa(s; 0)$ that a trajectory-horn β_σ associated with σ is expressed as

$$\beta_\sigma(s, t) = \frac{1}{2\kappa}\big\{\sin\kappa t - \sqrt{-1}(\cos\kappa t - 1)\big\}\big(\sqrt{4 - \kappa^2 s^2} + \sqrt{-1}|\kappa|s\big)u.$$

We hence have $\mu_\kappa(s; 0) = \big\|\frac{\partial\beta_\sigma}{\partial s}\big(s, r_\kappa(s; 0)\big)\big\| = |\kappa|s/\sqrt{4 - \kappa^2 s^2}$. $\square$

Proposition 3.2. *On $\mathbb{C}P^n(c)$, for $0 \le s \le d(\kappa; c)$, we have the following:*

(1) *The tube-length satisfies the relation*
$$\sqrt{c}\sin\big(\sqrt{\kappa^2 + c}\, r_\kappa(s; c)/2\big) = \sqrt{\kappa^2 + c}\,\sin(\sqrt{c}\, s/2),$$
hence is monotone increasing;

(2) *The tube-cosine is given by*
$$\epsilon_\kappa(s; c) = \sqrt{1 - (\kappa^2/c)\tan^2\big(\sqrt{c}\, s/2\big)},$$
hence is monotone decreasing;

(3) *When $s < d(\kappa; c)$, the expansion is given by*
$$\mu_\kappa(s; c) = |\kappa|\tan\big(\sqrt{c}\, s/2\big)\big\{c - \kappa^2\tan^2\big(\sqrt{c}\, s/2\big)\big\}^{-1/2},$$
hence is monotone increasing and satisfies $\lim_{s\uparrow d(\kappa; c)}\mu_\kappa(s; c) = \infty$.

Proof. We first study the case $c = 4$. A horizontal lift $\hat\sigma$ of a geodesic σ with respect to the Hopf fibration $\varpi : S^{2n+1}\ (\subset \mathbb{C}^{n+1}) \to \mathbb{C}P^n(4)$ of a unit sphere is given as $\hat\sigma(s) = \cos s\, z + \sin s\, u$, where $z \in S^{2n+1} \subset \mathbb{C}^{n+1}$ satisfies $\varpi(z) = \sigma(0)$ and a horizontal vector $(z, u) \in T_z S^{2n+1} \subset \{z\} \times \mathbb{C}^{n+1}$ with respect to ϖ satisfies $d\varpi\big((z, u)\big) = \dot\sigma(0)$. As we see in [1], a horizontal lift $\hat\gamma$ of a trajectory γ for $\mathbb{B}_\kappa$ with $\gamma(0) = \sigma(0)$ is of the form

$$\hat\gamma(t) = e^{\sqrt{-1}\kappa t/2}\left\{\cos\frac{\sqrt{\kappa^2 + 4}\, t}{2}\, z + \frac{1}{\sqrt{\kappa^2 + 4}}\sin\frac{\sqrt{\kappa^2 + 4}\, t}{2}\big(-\sqrt{-1}\kappa z + 2v\big)\right\}.$$

Since σ and γ lie on a same totally geodesic $\mathbb{C}P^1$ we have $v = e^{\sqrt{-1}\theta}u$ with some real θ. If $\gamma(r_\kappa(s; 4)) = \sigma(s)$ holds, then there is a real number

ψ satisfying $\hat{\sigma}(s) = e^{\sqrt{-1}\psi}\hat{\gamma}(r_\kappa(s;4))$. We denote by $(\!(\ ,\ )\!)$ the canonical Hermitian product on $\mathbb{C}^{n+1}$. As $(\!(z,u)\!) = 0$, we have

$$\cos s = \exp\!\big(\sqrt{-1}\{\phi + (\kappa r_\kappa(s;4)/2)\}\big)$$
$$\times \left\{ \cos\frac{\sqrt{\kappa^2+4}\,r_\kappa(s;4)}{2} - \frac{\sqrt{-1}\kappa}{\sqrt{\kappa^2+4}}\sin\frac{\sqrt{\kappa^2+4}\,r_\kappa(s;4)}{2} \right\}, \quad (3.1)$$

$$\sin s = \exp\!\Big(\sqrt{-1}\{\phi+\theta+\tfrac{1}{2}\kappa r_\kappa(s;4)\}\Big)\frac{2}{\sqrt{\kappa^2+4}}\sin\frac{\sqrt{\kappa^2+4}\,r_\kappa(s;4)}{2}. \quad (3.2)$$

By (3.2) we have

$$\exp\!\Big(\sqrt{-1}\{\phi + \theta + \tfrac{1}{2}\kappa r_\kappa(s;4)\}\Big) = 1, \quad \sin s = \frac{2}{\sqrt{\kappa^2+4}}\sin\frac{\sqrt{\kappa^2+4}\,r_\kappa(s;4)}{2},$$

in particular, we see $\sin s \leq 2/\sqrt{\kappa^2+4}\ (<1)$. Hence by (3.1) we have

$$\cos^2 s = \cos^2\frac{\sqrt{\kappa^2+4}\,r_\kappa(s;4)}{2} + \frac{\kappa^2}{\kappa^2+4}\sin^2\frac{\sqrt{\kappa^2+4}\,r_\kappa(s;4)}{2}$$
$$= \cos^2\frac{\sqrt{\kappa^2+4}\,r_\kappa(s;4)}{2} + \frac{\kappa^2}{4}\sin^2 s.$$

As $\sin s \leq 2/\sqrt{\kappa^2+4}$ shows $0 \leq s \leq \tan^{-1}(2/|\kappa|)$, we find

$$e^{\sqrt{-1}\theta} = \frac{1}{\cos s}\left\{ \cos\frac{\sqrt{\kappa^2+4}\,r_\kappa(s;4)}{2} - \frac{\sqrt{-1}\kappa}{\sqrt{\kappa^2+4}}\sin\frac{\sqrt{\kappa^2+4}\,r_\kappa(s;4)}{2} \right\}$$
$$= \sqrt{1 - \frac{\kappa^2}{4}\tan^2 s} - \frac{\sqrt{-1}\kappa}{2}\tan s.$$

Considering the geodesic $\tau \mapsto \sigma(s-\tau)$ and the trajectory $t \mapsto \gamma(r_\kappa(s;4)-t)$ for $\mathbb{B}_{-\kappa}$, we find by their congruency to σ and γ through a same anti-holomorphic isometry that

$$\epsilon_\kappa(s;4) = \cos\theta = \sqrt{1 - (\kappa^2/4)\tan^2 s}.$$

A horizontal lift $\hat{\beta}_\sigma$ of a trajectory-horn β_σ associated with σ is expressed as

$$\hat{\beta}_\sigma(s,t) = e^{\sqrt{-1}\kappa t/2}\left\{ \left(\cos\frac{\sqrt{\kappa^2+4}\,t}{2} - \frac{\sqrt{-1}\kappa}{\sqrt{\kappa^2+4}}\sin\frac{\sqrt{\kappa^2+4}\,t}{2} \right)z \right.$$
$$\left. + \frac{2}{\sqrt{\kappa^2+4}}\sin\frac{\sqrt{\kappa^2+4}\,t}{2}\left(\sqrt{1 - \frac{\kappa^2}{4}\tan^2 s} - \frac{\sqrt{-1}\kappa}{2}\tan s \right)u \right\}.$$

Since $\|\hat{\beta}_\sigma(s,t)\| = 1$ we see that a horizontal lift of a magnetic Jacobi field $\frac{\partial \beta_\nu}{\partial s}$ is given by $\frac{\partial \hat{\beta}_\sigma}{\partial s} - ((\hat{\beta}_\sigma, \frac{\partial \hat{\beta}_\sigma}{\partial s}))\hat{\beta}_\sigma$. As we have

$$\frac{\partial \hat{\beta}_\sigma}{\partial s}(s, r_\kappa(s;4)) = -\exp\left(\frac{\sqrt{-1}}{2}\kappa r_\kappa(s;4)\right)\frac{\kappa \tan s}{2\cos s}\left(\frac{\kappa \tan s}{\sqrt{4 - \kappa^2 \tan^2 s}} + \sqrt{-1}\right)u,$$

we obtain $\mu_\kappa(s;4) = |\kappa| \tan s/\sqrt{4 - \kappa^2 \tan^2 s}$.

Generally, if we change the metric $\langle\,,\,\rangle$ on a Kähler manifold M homothetically to the metric $\lambda^2\langle\,,\,\rangle$ with some positive λ, then given a trajectory γ for $\mathbb{B}_\kappa$ with respect to the original metric, we find that the curve $\tilde{\gamma}$ defined by $\tilde{\gamma}(t) = \gamma(t/\lambda)$ is a trajectory for $\mathbb{B}_{\kappa/\lambda}$ with respect to the new metric, and that sectional curvatures change λ^{-2}-times to the original sectional curvatures. For a trajectory-horn β_σ for $\mathbb{B}_\kappa$ associated with σ on $\mathbb{C}P^n(c)$, by considering a new metric $\lambda^2\langle\,,\,\rangle$ with $\lambda = \sqrt{c}/2$, we get a trajectory-horn $\tilde{\beta}_{\tilde{\sigma}}(\tilde{s},\tilde{t}) = \beta_\sigma(\tilde{s}/\lambda, \tilde{t}/\lambda)$ for $\mathbb{B}_{\kappa/\lambda}$ on $\mathbb{C}P^n(4)$. Thus, by noticing $\frac{\partial \tilde{\beta}_{\tilde{\sigma}}}{\partial \tilde{s}}(\tilde{s},\tilde{t}) = \frac{1}{\lambda}\frac{\partial \beta_\sigma}{\partial s}(\tilde{s}/\lambda, \tilde{t}/\lambda)$ we have

$$r_\kappa(s;c) = r_{\kappa/\lambda}(\lambda s;4)/\lambda, \quad \epsilon_\kappa(s;c) = \epsilon_{\kappa/\lambda}(\lambda s;4), \quad \mu_\kappa(s;c) = \mu_{\kappa/\lambda}(\lambda s;4),$$

and get the conclusion. $\qquad\square$

Proposition 3.3. *On $\mathbb{C}H^n(c)$ we have the following for $0 \leq s \leq d(\kappa;c)$:*

(1) *The tube-length satisfies the relation*

$$\begin{cases} \sqrt{|c|}\sinh(\sqrt{|c| - \kappa^2}\, r_\kappa(s;c)/2) \\ \quad = \sqrt{|c| - \kappa^2}\sinh(\sqrt{|c|}\, s/2), & \text{when } 0 < |\kappa| < \sqrt{|c|}, \\[2ex] \sqrt{|c|}\, r_\kappa(s;c) = 2\sinh(\sqrt{|c|}\, s/2), & \text{when } \kappa = \pm\sqrt{|c|}, \\[2ex] \sqrt{|c|}\sin(\sqrt{\kappa^2 + c}\, r_\kappa(s;c)/2) \\ \quad = \sqrt{\kappa^2 + c}\sinh(\sqrt{|c|}\, s/2), & \text{when } |\kappa| > \sqrt{|c|}; \end{cases}$$

Hence it is monotone increasing, and when $|\kappa| \leq \sqrt{|c|}$ it satisfies $\lim_{s \to \infty} r_\kappa(s;c) = \infty$;

(2) *The tube-cosine is given by*

$$\epsilon_\kappa(s;c) = \sqrt{1 + (\kappa^2/c)\tanh^2(\sqrt{|c|}\, s/2)},$$

hence is monotone decreasing;

(3) *When $s < d(\kappa;c)$, the expansion is given by*

$$\mu_\kappa(s;c) = |\kappa|\tanh(\sqrt{|c|}\, s/2)\{|c| - \kappa^2 \tanh^2(\sqrt{|c|}\, s/2)\}^{-1/2},$$

hence is monotone increasing and satisfies $\lim_{s \uparrow d(\kappa;c)} \mu_\kappa(s;c) = \infty$.

Proof. When $c = -4$, a horizontal lift $\hat{\sigma}$ of a geodesic σ with respect to the fibration $\varpi : H_1^{2n+1} (\subset \mathbb{C}_1^{n+1}) \to \mathbb{C}H^n(-4)$ of an anti-de Sitter space H_1^{2n+1} is given as $\hat{\sigma}(s) = \cos s\, z + \sin s\, u$, where $z \in H_1^{2n+1} \subset \mathbb{C}^{n+1}$ satisfies $\varpi(z) = \sigma(0)$ and a horizontal vector $(z, u) \in T_z H_1^{2n+1} \subset \{z\} \times \mathbb{C}^{n+1}$ with respect to ϖ satisfies $d\varpi\big((z, u)\big) = \dot{\sigma}(0)$. Here, the Hermitian product $\langle\!\langle\ ,\ \rangle\!\rangle$ on $\mathbb{C}_1^{n+1}$ is give by $\langle\!\langle z, w \rangle\!\rangle = -z_0 \bar{w}_0 + z_1 \bar{w}_1 + \cdots + z_n \bar{w}_n$ for $z = (z_0, \ldots, z_n)$, $w = (w_0, \ldots, w_n) \in \mathbb{C}^{n+1}$. As we see in [1], a horizontal lift $\hat{\gamma}$ of a trajectory γ for $\mathbb{B}_\kappa$ with $\gamma(0) = \sigma(0)$ is of the form

$$
\hat{\gamma}(t) = \begin{cases} e^{\sqrt{-1}\kappa t/2}\left\{\cosh \dfrac{\sqrt{4-\kappa^2}\,t}{2}\, z + \dfrac{1}{\sqrt{4-\kappa^2}} \sinh \dfrac{\sqrt{4-\kappa^2}\,t}{2}\left(-\sqrt{-1}\kappa z + 2v\right)\right\}, \\[4pt] \qquad\qquad\qquad\qquad\qquad\qquad\qquad\qquad\qquad\qquad \text{when } |\kappa| < 2, \\[8pt] e^{\pm\sqrt{-1}t}\left\{(1 \mp \sqrt{-1}\,t)z + tv\right\}, \qquad\qquad\quad \text{when } \kappa = \pm 2, \\[8pt] e^{\sqrt{-1}\kappa t/2}\left\{\cos \dfrac{\sqrt{\kappa^2-4}\,t}{2}\, z + \dfrac{1}{\sqrt{\kappa^2-4}} \sin \dfrac{\sqrt{\kappa^2-4}\,t}{2}\left(-\sqrt{-1}\kappa z + 2v\right)\right\}, \\[4pt] \qquad\qquad\qquad\qquad\qquad\qquad\qquad\qquad\qquad\qquad \text{when } |\kappa| > 2. \end{cases}
$$

Since σ and γ lie on a same totally geodesic $\mathbb{C}H^1$ we have $v = e^{\sqrt{-1}\theta} u$ with some real θ. If $\gamma(r_\kappa(s; -4)) = \sigma(s)$ holds, then there is a real number ψ satisfying $\hat{\sigma}(s) = e^{\sqrt{-1}\psi} \hat{\gamma}(r_\kappa(s; -4))$. As z and u are linearly independent over $\mathbb{C}$, by comparing their coefficients we have $e^{\sqrt{-1}\{\phi+\theta+(\kappa r_\kappa(s;-4)/2)\}} = 1$ and

$$
\sinh s = \begin{cases} \dfrac{2}{\sqrt{4-\kappa^2}} \sinh\left(\dfrac{\sqrt{4-\kappa^2}\,r_\kappa(s; -4)}{2}\right), & \text{if } |\kappa| < 2, \\[8pt] r_\kappa(s; -4), & \text{if } \kappa = \pm 2, \\[8pt] \dfrac{2}{\sqrt{\kappa^2-4}} \sin\left(\dfrac{\sqrt{\kappa^2-4}\,r_\kappa(s; -4)}{2}\right), & \text{if } |\kappa| > 2, \end{cases}
$$

$e^{\sqrt{-1}\theta} \cosh s$

$$
= \begin{cases} \cosh\left(\dfrac{\sqrt{4-\kappa^2}\,r_\kappa(s;-4)}{2}\right) - \dfrac{\sqrt{-1}\,\kappa}{\sqrt{4-\kappa^2}} \sinh\left(\dfrac{\sqrt{4-\kappa^2}\,r_\kappa(s;-4)}{2}\right), & \text{if } |\kappa| < 2, \\[8pt] 1 \mp \sqrt{-1}\,r_\kappa(s; -4), & \text{if } \kappa = \pm 2, \\[8pt] \cos\left(\dfrac{\sqrt{\kappa^2-4}\,r_\kappa(s;-4)}{2}\right) - \dfrac{\sqrt{-1}\,\kappa}{\sqrt{\kappa^2-4}} \sin\left(\dfrac{\sqrt{\kappa^2-4}\,r_\kappa(s;-4)}{2}\right), & \text{if } |\kappa| > 2. \end{cases}
$$

Considering the geodesic $\tau \mapsto \sigma(s-\tau)$ and the trajectory $t \mapsto \gamma\big(r_\kappa(s; -4) - t\big)$ for $\mathbb{B}_{-\kappa}$ we find

$$
\epsilon_\kappa(s; -4) = \cos\theta = \sqrt{1 - (\kappa^2/4)\tanh^2 s}.
$$

A horizontal lift $\hat{\beta}_\sigma$ of a trajectory-horn β_σ for $\mathbb{B}_\kappa$ associated with σ is expressed as

$$\hat{\beta}_\sigma(s,t) = e^{\sqrt{-1}\kappa t/2}\left\{\left(\cosh\frac{\sqrt{4-\kappa^2}\,t}{2} - \frac{\sqrt{-1}\kappa}{\sqrt{4-\kappa^2}}\sinh\frac{\sqrt{4-\kappa^2}\,t}{2}\right)z\right.$$

$$\left. + \frac{2}{\sqrt{4-\kappa^2}}\sinh\frac{\sqrt{4-\kappa^2}\,t}{2}\left(\sqrt{1-\frac{\kappa^2}{4}\tanh^2 s} - \frac{\sqrt{-1}\kappa}{2}\tanh s\right)u\right\}$$

when $|\kappa| < 2$,

$$\hat{\beta}_\sigma(s,t) = e^{\pm\sqrt{-1}t}\left\{(1\mp\sqrt{-1}\,t)z + t\left(\frac{1}{\cosh s}\mp\sqrt{-1}\tanh s\right)u\right\}$$

when $\kappa = \pm 2$, and

$$\hat{\beta}_\sigma(s,t) = e^{\sqrt{-1}\kappa t/2}\left\{\left(\cos\frac{\sqrt{\kappa^2+4}\,t}{2} - \frac{\sqrt{-1}\kappa}{\sqrt{\kappa^2+4}}\sin\frac{\sqrt{\kappa^2+4}\,t}{2}\right)z\right.$$

$$\left. + \frac{2}{\sqrt{\kappa^2+4}}\sin\frac{\sqrt{\kappa^2+4}\,t}{2}\left(\sqrt{1-\frac{\kappa^2}{4}\tan^2 s} - \frac{\sqrt{-1}\kappa}{2}\tan s\right)u\right\}$$

when $|\kappa| > 2$. Since $\langle\!\langle\hat{\beta}_\sigma,\hat{\beta}_\sigma\rangle\!\rangle = -1$, we see that a horizontal lift of a magnetic Jacobi field $\frac{\partial\beta_\sigma}{\partial s}$ is given by $\frac{\partial\hat{\beta}_\sigma}{\partial s} + \langle\!\langle\frac{\partial\hat{\beta}_\sigma}{\partial s},\hat{\beta}_\sigma\rangle\!\rangle\hat{\beta}_\sigma$. As we have

$$\frac{\partial\hat{\beta}_\sigma}{\partial s}\big(s,r_\kappa(s;-4)\big)$$

$$= \begin{cases} -e^{\sqrt{-1}\kappa r_\kappa(s;-4)/2}\dfrac{\kappa\tanh s}{2\cosh s}\left(\dfrac{\kappa\tanh^2 s}{\sqrt{4-\kappa^2\tanh^2 s}} + \sqrt{-1}\right)u, & \text{if } |\kappa| < 2, \\[3ex] -e^{\sqrt{-1}\sinh s}\dfrac{\tanh s}{\cosh s}\big(\sinh s \pm \sqrt{-1}\big)u, & \text{if } \kappa = \pm 2, \\[3ex] e^{\sqrt{-1}\kappa r_\kappa(s;-4)/2}\dfrac{\kappa\tanh s}{2\cosh s}\left(\dfrac{\kappa\tanh^2 s}{\sqrt{4-\kappa^2\tanh^2 s}} - \sqrt{-1}\right)u, & \text{if } |\kappa| > 2 \end{cases}$$

for $0 \le s < d(\kappa;-4)$, we have $\mu_\kappa(s;-4) = |\kappa|\tanh s/\sqrt{4-\kappa^2\tanh^2 s}$. Considering homothetical change of the metric we get the conclusion. □

Remark 3.1. Since two trajectories for $\mathbb{B}_\kappa$ on $\mathbb{C}M^n(c)$ are congruent to each other, a trajectory-ball of radius $r_\kappa(s;c)$ coincides with a geodesic ball of radius s. Hence the vector $Z_{\sigma,\kappa}(s) = \frac{\partial\beta_\sigma}{\partial s}\big(s,r_\kappa(s;c)\big)$ for a trajectory-horn β_σ is orthogonal to $\dot{\sigma}(s)$. We hence have $dr_\kappa(s;c)/ds = 1/\epsilon_\kappa(s;c)$.

We here summarize some properties of tube-lengths, tube-cosines and expansions of trajectory-horns on $\mathbb{C}M^n(c)$. We can get them by direct computation.

Lemma 3.1. *For appropriate s we have the following:*

(1) *When* $|\kappa_1| < |\kappa_2|$,

$$r_{\kappa_1}(s;c) < r_{\kappa_2}(s;c), \quad \epsilon_{\kappa_1}(s;c) > \epsilon_{\kappa_2}(s;c), \quad \mu_{\kappa_1}(s;c) < \mu_{\kappa_2}(s;c);$$

(2) *When* $c_1 < c_2$,

$$r_\kappa(s;c_1) < r_\kappa(s;c_2), \quad \epsilon_\kappa(s;c_1) > \epsilon_\kappa(s;c_2), \quad \mu_\kappa(s;c_1) < \mu_\kappa(s;c_2).$$

4. Tube-lengths and tube-cosines for trajectory-horns

In this section we study tube-lengths and tube-cosines for trajectory-horns on general Kähler manifolds by comparing them to those on complex space forms. For a trajectory-horn β_σ for $\mathbb{B}_\kappa$ associated with a geodesic $\sigma : [0, S] \to M$ and for a constant c, we set $S_{\sigma,\kappa}(c)$ the minimum positive s_* satisfying $r_{\sigma,\kappa}(s_*) = d(\kappa; c)$. In the case that there are no such s_* we set $S_{\sigma,\kappa}(c) = S$. Similarly, we set a positive number $q_{\sigma,\kappa}$ as the minimum positive $s_\sharp$ satisfying $r_{\sigma,\kappa}(s_\sharp) = \iota_{\sigma(0)}$. In the case that there are no such $s_\sharp$ we set $q_{\sigma,\kappa} = S$. For $0 < s \leq q_{\sigma,\kappa}$, as we have $d(\sigma(0), \sigma(s)) < r_{\sigma,\kappa}(s) \leq \iota_{\sigma(0)}$, we see that σ is the minimal geodesic joining $\sigma(0)$ and $\sigma(s)$, in particular, we have $d(\sigma(0), \sigma(s)) = s$. For s_0 with $0 < s_0 \leq S$, we set

$$\mathcal{HR}_{\sigma,\kappa}(s_0) = \{\beta_\sigma(s,t) \mid 0 \leq s \leq s_0, \, 0 \leq t \leq r_{\sigma,\kappa}(s)\}$$

and call it the horn-body of this trajectory-horn at s_0. When sectional curvatures are bounded from above, we have the following estimates on tube-lengths and tube-cosines.

Theorem 4.1. *Let β_σ be a trajectory-horn for $\mathbb{B}_\kappa$ associated with a geodesic $\sigma : [0, S] \to M$ on a Kähler manifold M whose sectional curvatures satisfy $\mathrm{Riem}^M \leq c$ with a constant c. Then its tube-length and its tube-cosine satisfy*

$$s < r_{\sigma,\kappa}(s) \leq r_\kappa(s;c), \quad \epsilon_{\sigma,\kappa}(s) \geq \epsilon_\kappa(s;c)$$

for $0 \leq s \leq \min\{S_{\sigma,\kappa}(c), q_{\sigma,\kappa}\}$.

Moreover, if $r_{\sigma,\kappa}(s_0) = r_\kappa(s_0;c)$ holds at some s_0 with $0 < s_0 \leq \min\{S_{\sigma,\kappa}(c), q_{\sigma,\kappa}\}$, then the horn-body $\mathcal{HR}_{\sigma,\kappa}(s_0)$ is totally geodesic, holomorphic and of constant sectional curvature c. In particular, $r_{\sigma,\kappa}(s) = r_\kappa(s;c)$ and $\epsilon_{\sigma,\kappa}(s) = \epsilon_\kappa(s;c)$ hold for $0 \leq s \leq s_0$.

Remark 4.1. When M is a Hadamard manifold, a simply connected manifold of nonpositive curvature, we have $q_{\sigma,\kappa} = \infty$ for each geodesic half-line σ and an arbitrary κ. If this manifold satisfies $\mathrm{Riem}^M \leq C < 0$ and if $|\kappa| \leq \sqrt{|C|}$, then for each geodesic half-line σ we have $S_{\sigma,\kappa}(c) = \infty$ for $c \leq C$ because $d(\kappa; c) = \infty$. In this sense, our comparison theorems on trajectory-horns are useful for a Hadamard Kähler manifold satisfying $\mathrm{Riem}^M \leq C$ and a Kähler magnetic field $\mathbb{B}_\kappa$ with $|\kappa| \leq \sqrt{|C|}$.

We shall show Theorem 4.1 by support of comparison theorems on trajectory-harps. Let $\gamma : [0,T] \to M$ be a trajectory-segment or a trajectory-half line for a non-trivial $\mathbb{B}_\kappa$ satisfying $\gamma(t) \neq \gamma(0)$ for $0 < t < T$. A smooth variation $\alpha_\gamma : [0,T] \times \mathbb{R} \to M$ of geodesics is said to be a *trajectory-harp* associated with γ if it satisfies

i) $\alpha_\gamma(t,0) = \gamma(0)$ for $0 \leq t \leq T$,

ii) when $t = 0$, the curve $s \mapsto \alpha_\gamma(0,s)$ is the geodesic of initial vector $\dot\gamma(0)$,

iii) when $0 < t < T$ the curve $s \mapsto \alpha_\gamma(t,s)$ is a geodesic of unit speed joining $\gamma(0)$ and $\gamma(t)$.

When the image $\gamma([0,T])$ of a trajectory γ is contained in the geodesic ball $B_{\iota_p}(p)$ centered at $p = \gamma(0)$ and of radius ι_p of injectivity at p, as we can join $\gamma(0)$ and $\gamma(t)$ by a unique minimizing geodesic, we have a unique trajectory-harp. Therefore, if the image $\gamma([0,T])$ lies in the geodesic ball $B_{c_p}(p)$ centered at $p = \gamma(0)$ and of radius c_p of first conjugate value of p, we have a trajectory-harp by extending the previous one.

For a trajectory-harp α_γ, we call the geodesic segment $s \mapsto \alpha_\gamma(t,s)$ joining $\gamma(0)$ and $\gamma(t)$ the *string* of this trajectory-harp at t. We denote by $\ell_\gamma(t)$ the length of this string, set $\ell_\gamma(0) = 0$, and call it the string-length at t. We set $\delta_\gamma(t) = \langle \dot\gamma, \frac{\partial \alpha_\gamma}{\partial s} \rangle$ and call it the string-cosine at t. It is known that the derivative of the string-length is the string-cosine (i.e. $\ell'_\gamma(t) = \delta_\gamma(t)$, see [5]). We put $R_\gamma = \sup\{t \mid \delta_\gamma(\tau) > 0$ for $0 < \tau < t\}$ and call it the maximal arch length of a trajectory-harp α_γ. For constants κ and c, we denote by $\ell_\kappa(\cdot, c) : [0, \pi/\sqrt{\kappa^2 + c}] \to \mathbb{R}$ the inverse function of $r_\kappa(\cdot; c) : [0, d(\kappa; c)] \to \mathbb{R}$, and set $\delta_\kappa(t; c) = \epsilon_\kappa(\ell_\kappa(t; c); c)$. We set $T_\gamma(c)$ as the minimum positive t_* satisfying $\ell_\gamma(t_*) = d(\kappa; c) \left(= \ell_\kappa(\pi/\sqrt{\kappa^2 + c}) \right)$. In case we do not have such t_* we set $T_\gamma(c) = T$. For $0 < t_0 \leq T$, we put $\mathcal{HP}_\gamma(t_0) = \{\alpha_\gamma(t,s) \mid 0 \leq t \leq t_0, 0 \leq s \leq \ell_\gamma(t)\}$ and call it the harp-body of α_γ at t_0.

Lemma 4.1. *For a trajectory $\gamma : [0,T] \to M$ for $\mathbb{B}_\kappa$ and for a constant c, we have $T_\gamma(c) \geq d(\kappa; c)$ if $T \geq d(\kappa; c)$.*

Proof. We take $0 \leq t < d(\kappa; c)$. As we have $\ell_\gamma(t) \leq t$, we find

$$\ell_\kappa\left(\pi/\sqrt{\kappa^2 + c}; c\right) = d(\kappa; c) > t \geq \ell_\gamma(t),$$

which shows the conclusion.   □

Proposition 4.1 ([5, 7]). *Let α_γ be a trajectory-harp associated with a trajectory $\gamma : [0, T] \to M$ for $\mathbb{B}_\kappa$ on a Kähler manifold M. Suppose that sectional curvatures of planes tangent to the harp-body $\mathcal{HP}_\gamma(T)$ are not greater than a constant c. We then have the following:*

1) $\ell_\gamma(t) \geq \ell_\kappa(t; c)$ *for* $0 \leq t \leq \min\{R_\gamma, 2\pi/\sqrt{\kappa^2 + c}\}$,
2) $\delta_\gamma(t) \geq \delta_\kappa\big(r_\kappa(\ell_\gamma(t); c); c\big)$ *for* $0 \leq t \leq T_\gamma(c)$, *in particular*, $R_\gamma \geq T_\gamma(c)$.

Moreover, if $\ell_\gamma(t_0) = \ell_\kappa(t_0; c)$ *holds at some t_0 with $0 < t_0 \leq \min\{R_\gamma, \pi/\sqrt{\kappa^2 + c}\}$, then the harp-body $\mathcal{HP}_\gamma(t_0)$ is totally geodesic, holomorphic and of constant sectional curvature c. In particular, we have* $\ell_\gamma(t) = \ell_\kappa(t; c)$ *and* $\delta_\gamma(t) = \delta_\kappa(t; c)$ *for* $0 \leq t \leq t_0$.

We shall show Theorem 4.1 by using the above comparison theorem on trajectory-harps.

Proof of Theorem 4.1. We fix s and consider a trajectory-segment γ_s given by $\gamma_s(t) = \beta_\sigma(s, t)$ for $0 \leq t \leq r_{\sigma,\kappa}(s)$. Since the distance between $\sigma(0)$ and $\beta_{\sigma,\kappa}(s, t)$ satisfies $d\big(\sigma(0), \beta_{\sigma,\kappa}(s, t)\big) < r_{\sigma,\kappa}(s) \leq \iota_{\sigma(0)}$ for $0 \leq s \leq q_{\sigma,\kappa}$ and for $0 \leq t \leq r_{\sigma,\kappa}(s)$, we can construct a trajectory-harp α_{γ_s} associated with γ_s so that each string is a minimizing geodesic. As we see before that for $0 < s \leq q_{\sigma,\kappa}$ the geodesic segment $\sigma|_{[0,s]}$ is the minimizing geodesic joining $\sigma(0)$ and $\sigma(s)$, we find that the string of α_{γ_s} at $r_{\sigma,\kappa}(s)$ coincides with this geodesic segment. We hence have $\ell_{\gamma_s}\big(r_{\sigma,\kappa}(s)\big) = s$. Since $r_{\sigma,\kappa}(s) \leq d(\kappa; c) \leq T_{\gamma_s}(c) \leq R_{\gamma_s}$, by Proposition 4.1 we have

$$\ell_\kappa\big(r_\kappa(s; c)\big) = s = \ell_{\gamma_s}\big(r_{\sigma,\kappa}(s)\big) \geq \ell_\kappa\big(r_{\sigma,\kappa}(s); c\big).$$

As $\ell_\kappa(\cdot; c)$ is monotone increasing, we get the first inequality.

Since the string of α_{γ_s} at $r_{\sigma,\kappa}(s)$ coincides with $\sigma|_{[0,s]}$ and as $r_{\sigma,\kappa}(s) \leq T_{\gamma_s}(c)$, by Proposition 4.1 we have

$$\epsilon_{\sigma,\kappa}(s) = \delta_{\gamma_s}\big(r_{\sigma,\kappa}(s)\big)$$
$$\geq \delta_\kappa\big(r_\kappa(\ell_{\gamma_s}(r_{\sigma,\kappa}(s)); c); c\big) = \delta_\kappa\big(r_\kappa(s; c); c\big) = \epsilon_\kappa(s; c).$$

We now suppose $r_{\sigma,\kappa}(s_0) = r_\kappa(s_0; c)$. Considering the trajectory-harp $\alpha_{\gamma_{s_0}}$ associated with $\gamma_{s_0} : [0, r_{\sigma,\kappa}(s_0)] \to M$, as we have $\ell_{\gamma_{s_0}}\big(r_{\sigma,\kappa}(s_0)\big) = s_0 = \ell_\kappa\big(r_\kappa(s_0; c); c\big)$, we find that the harp-body $\mathcal{HP}_{\gamma_{s_0}}\big(r_\kappa(s_0; c)\big)$ is totally geodesic, complex and of constant sectional curvature c. In particular, trajectories joining points $\sigma(0) = \gamma_{s_0}(0)$ and $\sigma(s)$ with $0 < s \leq s_0$ lie on this harp-body. Therefore, as $r_{\sigma,\kappa}(s) \leq r_\kappa(s; c) \leq \pi/\sqrt{\kappa^2 + c}$, we find that the horn-body $\mathcal{HR}_\sigma(s_0)$ is contained in the closure of a trajectory-ball of radius $c^\kappa_{\sigma(0)}$ centered at $\sigma(0)$. Hence, we see it coincides with that harp-body, and get the conclusion. □

As we make use of a comparison theorem on trajectory-harps to show our theorem, we need the condition that each tube lies in the closure of the geodesic ball $B_{\iota_{\sigma(0)}}(\sigma(0))$. Also, in the case that sectional curvatures of underlying manifold are bounded from below, to compare trajectory-harps we need additional assumption that they are holomorphic at their arches (see [12]). Therefore we can only estimate tube-lengths and tube-cosines of trajectory-horns on an orientable Riemann surface.

5. Expansions and embouchure angles for trajectory-horns

We here make mention of other quantities which show properties of trajectory-horns. For a trajectory-horn β_σ for $\mathbb{B}_\kappa$ associated with a geodesic σ, we consider the expansion vector field $\frac{\partial\beta_\sigma}{\partial s}(s, r_{\sigma,\kappa}(s))$ along σ. We note that $Y_s(t) = \frac{\partial\beta_\sigma}{\partial s}(s, t)$ is a magnetic Jacobi field along a trajectory $\gamma_s(t) = \beta_\sigma(s, t)$. As we see in Lemma 2.1, the expression $\sigma(s) = \beta_\sigma(s, r_{\sigma,\kappa}(s))$ leads us to $\dot{\sigma}(s) = Y_s(r_{\sigma,\kappa}(s)) + r'_{\sigma,\kappa}(s)\frac{\partial\beta_\sigma}{\partial t}(sr_{\sigma,\kappa}(s))$. By taking the inner products of both sides with $\dot{\gamma}_s(r_{\sigma,\kappa}(s))$ and by taking norms of both sides, we have

$$\epsilon_{\sigma,\kappa}(s) = \langle Y_s(r_{\sigma,\kappa}(s)), \dot{\gamma}_s(r_{\sigma,\kappa}(s))\rangle + r'_{\sigma,\kappa}(s),$$

$$1 = \left\|Y_s(r_{\sigma,\kappa}(s))\right\|^2 + 2r'_{\sigma,\kappa}(s)\langle Y_s(r_{\sigma,\kappa}(s)), \dot{\gamma}_s(r_{\sigma,\kappa}(s))\rangle + r'_{\sigma,\kappa}(s)^2.$$

By substituting the first equality to the second equality, we obtain $\left\|Y^\sharp_s(r_{\sigma,\kappa}(s))\right\|^2 = 1 - \epsilon_{\sigma,\kappa}(s)^2$. We set

$$\mu^\sharp_\kappa(s; c) = \sqrt{1 - \epsilon_\kappa(s; c)^2} = \begin{cases} (|\kappa|/\sqrt{c})\tan(\sqrt{c}\,s/2), & \text{when } c > 0, \\ |\kappa|\,s/2, & \text{when } c = 0, \\ (|\kappa|/\sqrt{|c|})\tanh(\sqrt{|c|}\,s/2), & \text{when } c < 0. \end{cases}$$

This is the norm of the component of the expansion vector field orthogonal to tubes for a trajectory-horn for $\mathbb{B}_\kappa$ on $\mathbb{C}M^n(c)$. By applying Theorem 4.1 we have the following.

Proposition 5.1. *Let β_σ be a trajectory-horn for $\mathbb{B}_\kappa$ associated with a geodesic $\sigma : [0, S] \to M$ on a Kähler manifold M whose sectional curvatures satisfy $\mathrm{Riem}^M \leq c$ with some constant c. We then have*

$$\|Y^\sharp_s(r_{\sigma,\kappa}(s))\| \leq \mu^\sharp_\kappa(s; c) \qquad for \quad 0 \leq s \leq \min\{S_{\sigma,\kappa}, q_{\sigma,\kappa}\},$$

In view of Lemma 2.1, an important quantity is the inner product of the magnetic Jacobi field Y_s and the velocity vector $\dot{\gamma}_s$ of the trajectory $t \mapsto \beta_\sigma(s,t)$. On $\mathbb{C}M^n(c)$ this inner product $\mu_\kappa^0(s;c) = \langle Y_s(r_\kappa(s;c)), \dot{\gamma}_s(r_\kappa(s;c)) \rangle$ is given as follows:

$$\mu_\kappa^0(s;c) = \begin{cases} -\dfrac{\operatorname{sgn}(\kappa)\kappa^2 \tan^2(\sqrt{c}\,s/2)}{\sqrt{c^2 - c\kappa^2 \tan^2(\sqrt{c}\,s/2)}}, & \text{when } c > 0, \\[1.5em] -\dfrac{\operatorname{sgn}(\kappa)\kappa^2 s^2}{2\sqrt{4 - \kappa^2 s^2}}, & \text{when } c = 0, \\[1.5em] -\dfrac{\operatorname{sgn}(\kappa)\kappa^2 \tanh^2(\sqrt{c}\,s/2)}{\sqrt{c^2 - |c|\kappa^2 \tanh^2(\sqrt{c}\,s/2)}}, & \text{when } c < 0. \end{cases}$$

Trivially it satisfies $\lim_{s \uparrow d(\kappa;c)} \mu_\kappa^0(s;c) = -\infty$. Thus, we can guess that the behavior $\mu_{\sigma,\kappa}^0(s) = \langle Y_s(r_{\sigma,\kappa}(s)), \dot{\gamma}_s(r_{\sigma,\kappa}(s)) \rangle$ shows the rate of exhausion of the radius of the image $\mathbb{B}_\kappa \exp_{\sigma(0)}(T_{\sigma(0)}M)$ in the direction of $\dot{\sigma}(0)$. But since $\mu_\kappa^0(s;c)^2 = \mu_\kappa(s;c)^2 + \epsilon_\kappa(s;c)^2 - 1$, Lemma 3.1 suggests us that this quantity is not so easy to treat by an ordinary method of comparing.

On a general Kähler manifold, we can only give a rough estimate of this quantity under the assumption that sectional curvatures are bounded from above and below by using comparison theorems on magnetic Jacobi fields. For a constant C, we define a function $\mathfrak{s}(t;C)$ by

$$\mathfrak{s}(t;C) = \begin{cases} (1/\sqrt{C})\sin\sqrt{C}\,t, & \text{when } C > 0, \\ t, & \text{when } C = 0, \\ (1/\sqrt{|C|})\sinh\sqrt{|C|}\,t, & \text{when } C < 0. \end{cases}$$

We here suppose M satisfies $c_1 \leq \operatorname{Riem}^M \leq c_2$. Since $Y_s(0) = 0$, comparison theorem on magnetic Jacobi fields given in [4] guarantees that

- $\|Y_s^\sharp(t)\| \geq \left\|(\nabla_{\dot{\gamma}} Y_s)(0)\right\| \mathfrak{s}(t;\kappa^2 + c_2)$ for $0 \leq t \leq \pi/\sqrt{\kappa^2 + c_2}$ $(\leq c_{p,\gamma_s}^\kappa)$,
- $\|Y_s^\sharp(t)\| \leq \left\|(\nabla_{\dot{\gamma}} Y_s)(0)\right\| \mathfrak{s}\left(t;(\kappa^2/4) + c_1\right)$ for $0 \leq t \leq c_{p,\gamma_s}^\kappa$.

We note $\|Y_s^\sharp(t)\|^2 = 1 - \epsilon_{\sigma,\kappa}(s)^2 \leq 1$. As the second equality in (2.1) shows that $f_{Y_s}' = \kappa g_{Y_s}$ and as $|g_{Y_s}(t)| \leq \|Y_s^\sharp(t)\|$, we have

$$|\mu_{\sigma,\kappa}^0(s)| = \left| f_{Y_s}(r_{\sigma,\kappa}(s)) \right| \leq |\kappa| \int_0^{r_{\sigma,\kappa}(s)} |g_{Y_s}(t)|\, dt$$

$$\leq |\kappa| \int_0^{r_{\sigma,\kappa}(s)} \left\|(\nabla_{\dot{\gamma}} Y_s)(0)\right\| \mathfrak{s}\left(t; \frac{\kappa^2}{4} + c_1\right) dt$$

$$\leq \frac{|\kappa|\sqrt{1-\epsilon_{\sigma,\kappa}(s)^2}}{\mathfrak{s}\big(r_{\sigma,\kappa}(s);\kappa^2+c_2\big)} \int_0^{r_{\sigma,\kappa}(s)} \mathfrak{s}\Big(t;\frac{\kappa^2}{4}+c_1\Big)\,dt$$

if $r_{\sigma,\kappa}(s) \leq \pi/\sqrt{\kappa^2+c_2}$. In particular, $\mu_{\sigma,\kappa}^0(s)$ is finite in this interval. If we apply Theorem 4.1, we have $\epsilon_{\sigma,\kappa}(s) \geq \epsilon_{\kappa}(s;c_2)$ and $r_{\sigma,\kappa}(s) \leq r_{\kappa}(s;c_2)$ for $0 \leq s \leq \min\{S_{\sigma,\kappa}(c_2), q_{\sigma,\kappa}\}$, we hence obtain

$$|\mu_{\sigma,\kappa}^0(s)| \leq \frac{|\kappa|\,\mu_{\kappa}^{\sharp}(s;c_2)}{\mathfrak{s}\big(r_{\kappa}(s;c_1);\kappa^2+c_2\big)} \int_0^{r_{\kappa}(s;c_2)} \mathfrak{s}\Big(t;\frac{\kappa^2}{4}+c_1\Big)\,dt$$

for $0 \leq s \leq \min\{S_{\sigma,\kappa}(c_2), q_{\sigma,\kappa}\}$.

We next study another quantity of a trajectory-horn. For a trajectory-horn β_σ for $\mathbb{B}_\kappa$ associated with a geodesic $\sigma : [0,S] \to M$ and constants a, b with $0 \leq a < b \leq S$, we call the restriction $\beta_\sigma|_{[a,b]\times\mathbb{R}}$ a subhorn of this trajectory-horn. The length $\Theta_{\sigma,\kappa}(a,b)$ of the curve $[a,b] \ni s \mapsto \frac{\partial\beta_\sigma}{\partial t}(s,0) \in U_{\sigma(0)}M$ in the unit tangent space at $\sigma(0)$ is called the *embouchure angle* of this subhorn. Trivially the angle between two tubes are estimated as $\angle\big(\frac{\partial\beta_\sigma}{\partial t}(a,0), \frac{\partial\beta_\sigma}{\partial t}(b,0)\big) \leq \Theta_{\sigma,\kappa}(a,b)$. For a trajectory-horn for $\mathbb{B}_\kappa$ on $\mathbb{C}M^n(c)$, its embouchure angle $\Theta_\kappa(a,b;c)$ is the angle between two tubes and is given as

$$\Theta_\kappa(a,b;c) = \cos^{-1}\epsilon_\kappa(b;c) - \cos^{-1}\epsilon_\kappa(b;c) \qquad \text{for } 0 \leq a < b \leq S_\kappa(c).$$

We here give a rough estimate on embouchure angles.

Proposition 5.2. *Let β_σ be a trajectory horn for $\mathbb{B}_\kappa$ associated with a geodesic $\sigma : [0,S] \to M$ on a Kähler manifold M whose sectional curvatures satisfy $\mathrm{Riem}^M \leq c$ with a constant c. We then have*

$$\Theta_{\sigma,\kappa}(a,b) < \int_a^b \frac{\mu_\kappa^{\sharp}(s;c)}{\mathfrak{s}(s;\kappa^2+c)}\,ds$$

for $0 \leq a < b \leq \min\{S_{\sigma,\kappa}(c), q_{\sigma,\kappa}\}$.

Proof. As we see above, we have $\big\|Y_s^{\sharp}\big(r_{\sigma,\kappa}(s)\big)\big\|^2 = 1 - \epsilon_{\sigma,\kappa}(s)^2$ and have $\|Y_s^{\sharp}(t)\| \geq \big\|(\nabla_{\dot\gamma}Y)(0)\big\|\,\mathfrak{s}(t;\kappa^2+c)$ for $0 \leq t \leq \pi/\sqrt{\kappa^2+c}$ ([4]). By using Theorem 4.1, as the function $\mathfrak{s}(t;C)$ is monotone increasing with respect to t, we have

$$\Theta_{\sigma,\kappa}(a,b) = \int_a^b \big\|\big(\nabla_{\frac{\partial\beta_\sigma}{\partial s}}\frac{\partial\beta_\sigma}{\partial t}\big)(s,0)\big\|\,ds = \int_a^b \big\|\big(\nabla_{\frac{\partial\beta_\sigma}{\partial t}}Y_s\big)(0)\big\|\,ds$$

$$\leq \int_a^b \frac{\big\|Y_s^{\sharp}\big(r_{\sigma,\kappa}(s)\big)\big\|}{\mathfrak{s}\big(r_{\sigma,\kappa}(s);\kappa^2+c\big)}\,ds = \int_a^b \frac{\sqrt{1-\epsilon_{\sigma,\kappa}(s)^2}}{\mathfrak{s}\big(r_{\sigma,\kappa}(s);\kappa^2+c\big)}\,ds$$

$$\leq \int_a^b \frac{\sqrt{1 - \epsilon_\kappa(s;c)^2}}{\mathfrak{s}(r_{\sigma,\kappa}(s); \kappa^2 + c)}\, ds < \int_a^b \frac{\sqrt{1 - \epsilon_\kappa(s;c)^2}}{\mathfrak{s}(s; \kappa^2 + c)}\, ds$$

and get the conclusion. □

The author is interested in comparing embouchure angles of two sub-horns when their edge tubes correspondingly have the same lengths (cf. [3])

References

[1] T. Adachi, Kähler magnetic flows on a manifold of constant holomorphic sectional curvature, *Tokyo J. Math.* **18**, 473–483, (1995).

[2] ———, A comparison theorem for magnetic Jacobi fields, *Proc. Edinburgh Math. Soc.* **40**, 293–308, (1997).

[3] ———, A comparison theorem on sectors for Kähler magnetic fields, *Proc. Japan Acad. Sci.* **81** (**A**), 110–114, (2005).

[4] ———, Magnetic Jacobi fields for Kähler magnetic fields, in *Recent Progress in Differential Geometry and its Related Fields*, T. Adachi, H. Hashimoto & M. Hristov eds., World Scientific, Singapore 41–53, (2011).

[5] ———, A theorem of Hadamard-Cartan type for Kähler magnetic fields, *J. Math. Soc. Japan* **64**, 969–984, (2012).

[6] ———, A comparison theorem on harp-sectors for Kähler magnetic fields, *Southeast Asian Bull. Math.* **38**, 619–626, (2014).

[7] ———, Accurate trajectory-harps for Kähler magnetic fields, to appear in *J. Math. Soc. Japan.*

[8] J. Cheeger & D.G. Ebin, *Comparison Theorems in Riemannian Geometry*, Noth-Holland 1975, Amsterdam.

[9] S. Maeda & Y. Ohnita, Helical geodesic immersion into complex space forms, *Geom. Dedicata* **30**, 93–114, (1989).

[10] T. Sakai, *Riemannian Geometry*, Translations of monographs, A.M.S. 1996.

[11] Q. Shi & T. Adachi, Trajectory-harps and horns applied to the study of the ideal boundary of a Hadamard Kähler manifold, to appear in *Tokyo J. Math.*.

[12] ———, Comparison theorems on trajectory-harps for Kähler magnetic fields which are holomorphic at their arches, preprint (2016).

[13] T. Sunada, *Magnetic flows on a Riemann surface*, Proc. KAIST Math. Workshop 8(1993), 93–108.

Received November 15, 2016
Revised May 15, 2017

NULL CURVES ON THE UNIT TANGENT BUNDLE
OF A TWO-DIMENSIONAL KÄHLER-NORDEN MANIFOLD

In memory of Professor Todor Gramtchev,
University of Cagliari, Italy & Bulgarian Academy of Sciences

Galia NAKOVA*

Department of Algebra and Geometry, Faculty of Mathematics and Informatics,
University of Veliko Turnovo "St. Cyril end St. Metodius",
T. Tarnovski 2 str., 5003 Veliko Tarnovo, Bulgaria
E-mail: gnakova@gmail.com

In this paper we study null curves on the unit tangent bundle $U(TM)$ with a timelike unit normal vector field of a 2-dimensional Kähler-Norden manifold M. We find relationships between curves on the base manifold M and their corresponding null curves on $U(TM)$, which is a 3-dimensional Lorentzian manifold. The objects of investigation are also Legendre curves on $U(TM)$ with respect to the almost contact B-metric structure. We show that the considered Legendre null curves are Cartan framed null curves with respect to the original parameter. We also give some examples of such curves.

Keywords: Kähler-Norden manifolds; almost contact B-metric manifolds; Lorentzian manifolds; null curves; Legendre curves; Cartan framed null curves.

1. Introduction

A $2n$-dimensional manifold M is said to be a Kähler-Norden manifold if it is equipped with an almost complex structure J and a Norden metric g and if J is parallel with respect to the Levi-Civita connection of g. The Norden metric is a semi-Riemannian metric with a neutral signature and the almost complex structure is an anti-isometry with respect to the metric.

In this paper we consider a 2-dimensional Kähler-Norden manifold (M, J, g) that is a Lorentzian manifold. The tangent bundle TM of (M, J, g) is endowed with Sasaki metric $\bar{g}^S$, which is a semi-Riemannian metric on TM of signature $(2, 2)$. Then, the unit tangent bundle $U(TM)$ of TM with a timelike unit normal vector field is a 3-dimensional Lorentzian manifold with respect to the induced metric g' by $\bar{g}^S$. On $U(TM)$ we have an almost

---

*The author is partially supported by Scientific researches fund of "St. Cyril and St. Methodius" University of Veliko Tarnovo under contract RD-09-422-13 / 09.04.2014.

contact structure (φ, ξ, η) such that g' is a B-metric. Thus, $U(TM)$ becomes a 3-dimensional almost contact B-metric manifold $(U(TM), \varphi, \xi, \eta, g')$. A curve $C(t) = (\gamma(t), X(t))$ on $U(TM)$ which satisfies $g(\nabla_{\dot{\gamma}} X, \nabla_{\dot{\gamma}} X) > 0$ at an arbitrary t is said to be a curve of first type.

Our aim in the present paper is to study null curves of first type on the 3-dimensional Lorentzian manifold $(U(TM), g')$ and on the corresponding almost contact B-metric manifold $(U(TM), \varphi, \xi, \eta, g')$. Null curves on Lorentzian manifolds have important applications in general relativity. The general theory of null curves has been developed by Duggal, Bejancu, Jin in [2, 3].

We find that $C(t) = (\gamma(t), X(t))$ is a null curve of first type on $(U(TM), g')$ only when γ is a timelike curve on M satisfying a given condition. By the assumption that $\gamma(s)$ is a timelike curve parametrized by the arc length parameter s, we obtain a necessary and sufficient condition $C(s)$ to be a null curve of first type on $(U(TM), g')$ in terms of the function $\rho(s) = g(\gamma', J\gamma')$. We remark that the function ρ does not vanish in general, contrary to the case when the metric is of Hermitian type. For null curves of first type $C(s)$ on $(U(TM), \varphi, \xi, \eta, g')$, we find a necessary and sufficient condition $\eta(\dot{C})$ to be a constant in terms of the curvature of γ. Moreover, we determine the class of Legendre null curves of first type and show that they are non-geodesic.

One special feature of a null curve C is that a Frenet frame F along C and the Frenet formulas with respect to F are not unique. In [3] it is shown that if a non-geodesic null curve on a Lorentzian manifold is properly parametrized, then there exists a unique Frenet frame of C with minimum number of curvature functions in the corresponding Frenet formulas. Such a frame is called the Cartan Frenet frame, and a null curve together with its Cartan Frenet frame is called a Cartan framed null curve.

We show that the considered Legendre null curves of first type $C(s) = (\gamma(s), X(s))$ on $(U(TM), \varphi, \xi, \eta, g')$ are Cartan framed null curves with respect to the arc length parameter s of γ. Also, we construct examples of such curves.

2. Preliminaries

A $2n$-dimensional smooth manifold M is said to be *an almost complex manifold with Norden metric* if it is equipped with an almost complex structure J and a metric g such that J acts as an anti-isometry on each tangent space of M, that is,

$$J^2 X = -X, \quad g(JX, JY) = -g(X, Y),$$

for arbitrary vector fields X, Y on M. The metric g is semi-Riemannian of signature (n, n) and is known as Norden metric (or B-metric). An almost complex manifold with Norden metric is a Kähler-Norden manifold if $\nabla J = 0$, where ∇ is the Levi-Civita connection of g. From the condition $\nabla J = 0$, it follows that the curvature tensor R of type $(1, 3)$ of every Kähler-Norden manifold satisfies the equality $R(X, Y)JZ = JR(X, Y)Z$. Hence the curvature tensor R of type $(0, 4)$ defined by $R(X, Y, Z, W) = g(R(X, Y)Z, W)$ satisfies

$$R(X, Y, JZ, JW) = -R(X, Y, Z, W).$$

In the case when M is a 2-dimensional Kähler-Norden manifold, the above equality implies $R(e, Je, Je, e) = R(e, Je, e, Je)$, where $\{e, Je\}$ is a basis of $T_x M$ at an arbitrary point $x \in M$. Since $R(e, Je, Je, e) = -R(e, Je, e, Je)$ we obtain $R(e, Je, Je, e) = R(e, Je, e, Je) = 0$. Hence, all components of R are zero. We therefore find that every 2-dimensional Kähler-Norden manifold is flat.

Let $(M, \varphi, \xi, \eta, g')$ be a $(2n + 1)$-dimensional *almost contact B-metric manifold*, that is, (φ, ξ, η) is an almost contact structure and g' is a metric on M such that

$$\varphi^2 X = -X + \eta(X)\xi, \quad \eta(\xi) = 1,$$
$$g'(\varphi X, \varphi Y) = -g'(X, Y) + \eta(X)\eta(Y). \tag{1}$$

The metric g' is called a *B-metric* [4]. It is semi-Riemannian of signature $(n + 1, n)$. As immediate consequences from (1) we obtain

$$\eta \circ \varphi = 0, \quad \varphi\xi = 0, \quad \mathrm{rank}\,\varphi = 2n, \quad \eta(X) = g'(X, \xi), \quad g'(\xi, \xi) = 1.$$

Denote by TM the tangent bundle of an n-dimensional semi-Riemannian manifold (M, g). Recall that TM consists of pairs (x, u), where x is a point in M and u a tangent vector to M at x. Let $\pi : TM \to M$ given by $(x, u) \mapsto x$ be the natural projection of TM onto M and U an open subset of M. It is known that TM is a $2n$-dimensional smooth manifold. We denote a local coordinate system on $\pi^{-1}(U)$ by $(\overline{x}^1, \ldots, \overline{x}^n, \overline{u}^1, \ldots, \overline{u}^n)$. Here, when $(x^1, \ldots, x^n)$ is a local coordinate system on U and $(u^1, \ldots, u^n)$ are the coordinates of the tangent vector $u \in T_x M$ with respect to the local basis $\left\{\frac{\partial}{\partial x^i}\right\}$, $i = 1, \ldots, n$, then the coordinate functions $(\overline{x}^i; \overline{u}^i)$ on $\pi^{-1}(U)$ are determined as follows:

$$\overline{x}^i(x, u) = (x^i \circ \pi)(x, u) = x^i(x), \quad \overline{u}^i(x, u) = u(x^i) = u^i, \quad i = 1, \ldots, n.$$

It is known [1, 7] that the tangent space $T_{(x,u)}TM$ at an arbitrary point (x, u) splits into the following direct sum of subspaces

$$T_{(x,u)}TM = VTM_{(x,u)} \oplus HTM_{(x,u)},$$

where $VTM_{(x,u)} = \mathrm{Ker}\pi_{*|(x,u)}$ is the vertical subspace and $HTM_{(x,u)}$ is the horizontal subspace with respect to the Levi-Civita connection ∇ of g. For a function f on M, we can consider the 1-form df as a function $\overline{df}$ on TM by setting $\overline{df} = \left(\frac{\partial f}{\partial x^i} \circ \pi\right)\overline{u}^i$. Given a tangent vector X_x belonging to T_xM, we can define a unique vector X_x^v at the point (x, u) such that $X_x^v \in VTM_{(x,u)}$ and $X_x^v(\overline{df}) = X_x(f)$ for an arbitrary function f on M. We call it the *vertical lift* of X_x to (x, u). The unique vector X_x^h at the point (x, u) such that $X_x^h \in HTM_{(x,u)}$ and $\pi_*(X_x^h) = X_x$ is called the *horizontal lift* of X_x to (x, u). Both subspaces $VTM_{(x,u)}$ and $HTM_{(x,u)}$ are isomorphic to $T_{(x,u)}TM$. Hence, every tangent vector $\overline{Z} \in T_{(x,u)}TM$ can be decomposed as $\overline{Z} = X_x^h + Y_x^v$ with uniquely determined vectors $X_x, Y_x \in T_xM$. The *horizontal* and *vertical lifts* X^h, X^v of a vector field X on M to TM are the vector fields on TM which satisfy that X_x^h and X_x^v are the horizontal and vertical lifts of X_x at an arbitrary point, respectively.

With respect to the local coordinate system on $\pi^{-1}(U)$ the horizontal and vertical lifts of a vector field $X = X^i\frac{\partial}{\partial x^i}$ on U are defined by

$$X^h = (X^i \circ \pi)\frac{\partial}{\partial \overline{x}^i} - \overline{u}^b((X^a\Gamma_{ab}^i) \circ \pi)\frac{\partial}{\partial \overline{u}^i}, \tag{2}$$

$$X^v = (X^i \circ \pi)\frac{\partial}{\partial \overline{u}^i}, \tag{3}$$

where Γ_{ab}^i are the Christoffel symbols of ∇. For the horizontal and vertical lifts of $\frac{\partial}{\partial x^i}$ we have

$$\left(\frac{\partial}{\partial x^i}\right)^h = \frac{\partial}{\partial \overline{x}^i} - \overline{u}^b(\Gamma_{ib}^\alpha \circ \pi)\frac{\partial}{\partial \overline{u}^\alpha}, \qquad \left(\frac{\partial}{\partial x^i}\right)^v = \frac{\partial}{\partial \overline{u}^i}. \tag{4}$$

3. Null curves on the unit tangent bundle of a two-dimensional Kähler-Norden manifold

Let (M, J, g) be a 2-dimensional Kähler-Norden manifold. We consider the tangent bundle TM of M endowed with *Sasaki metric* $\overline{g}^S$ determined by

$$\overline{g}^S(X^h, Y^h) = \overline{g}^S(X^v, Y^v) = g(X, Y) \circ \pi, \qquad \overline{g}^S(X^h, Y^v) = 0.$$

Let $\{e_1, e_2\}$ be an orthonormal basis of T_xM such that $g(e_1, e_1) = -g(e_2, e_2) = -1$. Then $\{e_1^h, e_1^v, e_2^h, e_2^v\}$ is an orthonormal basis of $T_{(x,u)}TM$

with respect to $\overline{g}^S$ such that

$$\overline{g}^S(e_1^h, e_1^h) = \overline{g}^S(e_1^v, e_1^v) = -\overline{g}^S(e_2^h, e_2^h) = -\overline{g}^S(e_2^v, e_2^v) = -1.$$

Hence, $\overline{g}^S$ is a semi-Riemannian metric of signature $(2,2)$ on TM.

Denote by $\overline{\nabla}$ the Levi-Civita connection of $\overline{g}^S$. By using the formulas for $\overline{\nabla}$ of horizontal and vertical lifts of vector fields on a Riemannian manifold given in [1] and $R = 0$ on a 2-dimensional Kähler-Norden manifold M, we have

$$\overline{\nabla}_{X^v} Y^v = 0, \quad \overline{\nabla}_{X^v} Y^h = 0, \quad \overline{\nabla}_{X^h} Y^v = (\nabla_X Y)^v, \quad \overline{\nabla}_{X^h} Y^h = (\nabla_X Y)^h, \quad (5)$$

where X, Y are arbitrary vector fields on M.

Now, we define a unit tangent bundle $U(TM)$ of $(TM, \overline{g}^S)$ by

$$U(TM) = \{(x, u) \in TM : g(u, u) = -1\}. \quad (6)$$

We take the vertical vector field $\overline{N} = \overline{u}^i \frac{\partial}{\partial \overline{u}^i} = \overline{u}^i \left(\frac{\partial}{\partial x^i} \right)^v$ on TM. It is easy to verify that at each point (x, u) of TM the vector $\overline{N}_{(x,u)} \in T_{(x,u)}TM$ coincides with the vertical lift u^v of the vector $u \in T_x M$. Hence, $\overline{N}$ is a unit timelike vector field on TM which is the normal vector field of $U(TM)$. It is clear that every vector field X^h on TM is a tangent vector field of $U(TM)$, but a vector field X^v on $U(TM)$ is not tangent to $U(TM)$ in general.

If $\{e_1 = u, e_2\}$ is an orthonormal basis of $T_x M$, then $\{u^h, e_2^h, e_2^v\}$ is an orthonormal basis of $T_{(x,u)}U(TM)$. Hence, the metric g' on $U(TM)$ induced by $\overline{g}^S$ is a semi-Riemannian metric of signature $(1, 2)$, that is, $(U(TM), g')$ is a 3-dimensional Lorentzian manifold. We denote by ∇' the Levi-Civita connection of g'. The formulas of Gauss and Weingarten are given as

$$\overline{\nabla}_{X'} Y' = \nabla'_{X'} Y' - g'(A_{\overline{N}} X', Y')\overline{N}, \quad \overline{\nabla}_{X'} \overline{N} = -A_{\overline{N}} X' \quad (7)$$

for arbitrary vector fields X', Y' tangent to $U(TM)$.

Let $\gamma : I \to M$ be a smooth curve on M given locally by

$$\gamma(t) = (x_1(t), x_2(t)), \quad t \in I \subseteq \mathbb{R}$$

and X be a smooth vector field defined along γ. Then $C(t) = (\gamma(t), X(t))$, $t \in I$, is a smooth curve on $U(TM)$. We here prepare some formulas for $C(t)$ which we need. If $X(t) = X^i(t) \frac{\partial}{\partial x^i} \Big|_{\gamma(t)}$, for the tangent vector $\dot{C}(t)$ of $C(t)$ we have

$$\dot{C}(t) = \left(\frac{d\overline{x}^i}{dt} \frac{\partial}{\partial \overline{x}^i} + \frac{d(X^i \circ \pi)}{dt} \frac{\partial}{\partial \overline{u}^i} \right) (C(t)).$$

Taking into account (4) we obtain

$$\dot{C}(t)$$

$$= \left(\left(\frac{dx^i}{dt} \frac{\partial}{\partial x^i} \right)^h + \left[\frac{d(X^i \circ \pi)}{dt} + (X^b \circ \pi) \left(\frac{dx^a}{dt} \Gamma^i_{ab} \right) \circ \pi \right] \left(\frac{\partial}{\partial x^i} \right)^v \right) (C(t)) \quad (8)$$

$$= \left((\dot{\gamma})^h + (\nabla_{\dot{\gamma}} X)^v \right) (C(t)).$$

Remark 3.1.

(i) When $\gamma(t)$ is a *regular curve* (i.e. $\dot{\gamma}(t) \neq 0$ for every $t \in I$), the curve $C(t)$ is also regular by (8).

(ii) Since $\overline{N}(t) = X(t)^v$ at every point of $C(t)$, we have $\overline{N}_{|C} = X^v$.

(iii) The condition $g(X, X) = -1$ implies $g(\nabla_{\dot{\gamma}} X, X) = 0$. Then we have $\overline{g}^S(\overline{N}, (\nabla_{\dot{\gamma}} X)^v) = \overline{g}^S(X^v, (\nabla_{\dot{\gamma}} X)^v) = 0$. Hence, $(\nabla_{\dot{\gamma}} X)^v$ is a tangent vector field of $U(TM)$ defined along C.

Further, we assume that $\gamma(t)$ is a regular curve on M. Replacing $\overline{N}$ by X^v in the first formula in (7) we express

$$\nabla'_{\dot{C}} \dot{C} = \overline{\nabla}_{\dot{C}} \dot{C} - \overline{g}^S(\overline{\nabla}_{\dot{C}} X^v, \dot{C}) X^v.$$

By using (5) and (8) we find

$$\overline{\nabla}_{\dot{C}} \dot{C} = (\nabla_{\dot{\gamma}} \dot{\gamma})^h + (\nabla_{\dot{\gamma}} \nabla_{\dot{\gamma}} X)^v. \quad (9)$$

Now, using that $\overline{g}^S(X^v, \dot{C}) = 0$ and (9), we have

$$\overline{g}^S(\overline{\nabla}_{\dot{C}} X^v, \dot{C}) = -\overline{g}^S(X^v, \overline{\nabla}_{\dot{C}} \dot{C})$$
$$= -g(X, \nabla_{\dot{\gamma}} \nabla_{\dot{\gamma}} X) \circ \pi = g(\nabla_{\dot{\gamma}} X, \nabla_{\dot{\gamma}} X) \circ \pi. \quad (10)$$

Finally, substituting (9) and (10) in the equality for $\nabla'_{\dot{C}} \dot{C}$ we obtain

$$\nabla'_{\dot{C}} \dot{C} = (\nabla_{\dot{\gamma}} \dot{\gamma})^h + (\nabla_{\dot{\gamma}} \nabla_{\dot{\gamma}} X)^v - \left(g(\nabla_{\dot{\gamma}} X, \nabla_{\dot{\gamma}} X) \circ \pi \right) X^v. \quad (11)$$

We take an arbitrary curve $\gamma(t)$ on (M, J, g) and a unit timelike vector field X along γ. Then $g(X, X) = -1$ implies $g(\nabla_{\dot{\gamma}} X, X) = 0$. When $(\nabla_{\dot{\gamma}} X)(t) \neq 0$, the vectors $X(t)$ and $(\nabla_{\dot{\gamma}} X)(t)$ form an orthogonal basis of $T_{\gamma(t)} M$. Now, taking into account that the metric g has a signature $(1, 1)$ on M, that is, (M, g) is a 2-dimensional Lorentzian manifold, we conclude that $g((\nabla_{\dot{\gamma}} X)(t), (\nabla_{\dot{\gamma}} X)(t)) > 0$ at this point. We call $C(t) = (\gamma(t), X(t))$ on $U(TM)$ a *curve of first type* if it satisfies $g(\nabla_{\dot{\gamma}} X, \nabla_{\dot{\gamma}} X) > 0$ everywhere.

On a semi-Riemannian manifold there exist three types of regular curves; spacelike, timelike and null curves. A regular curve $\gamma(t)$ on M is said to be *spacelike* (resp. *timelike*, *null* (*lightlike*)) if at every point of γ we have $g(\dot\gamma, \dot\gamma) > 0$ (resp. $g(\dot\gamma, \dot\gamma) < 0$, $g(\dot\gamma, \dot\gamma) = 0$). It is known that a regular spacelike or timelike curve γ can be parametrized by its arc length parameter s, that is, $|\gamma'(s)| = 1$ for all s. We note that if γ is spacelike (resp. timelike), then $|\gamma'(s)| = \sqrt{g(\gamma'(s), \gamma'(s))}$ (resp. $|\gamma'(s)| = \sqrt{-g(\gamma'(s), \gamma'(s))}$). For null curves there is no sense in considering parametrization by their arc length parameters.

Our aim in this paper is to study null curves of first type on $U(TM)$.

Proposition 3.1. *A curve $C(t) = (\gamma(t), X(t))$ on $U(TM)$ is a null curve of first type if and only if the condition $g(\nabla_{\dot\gamma} X, \nabla_{\dot\gamma} X) = -g(\dot\gamma, \dot\gamma) > 0$ holds at every point of γ.*

Proof. From (8) we obtain $g'(\dot C, \dot C) = g(\dot\gamma, \dot\gamma) \circ \pi + g(\nabla_{\dot\gamma} X, \nabla_{\dot\gamma} X) \circ \pi$. The assertion is a direct consequence of the definition of curves of first type. □

By Proposition 3.1 it is clear that only timelike curves on M induce null curves of first type on $U(TM)$. For later use we give some results concerning timelike curves on (M, J, g). For a timelike curve $\gamma(s)$ on a 2-dimensional Lorentzian manifold which is parametrized by its arc length parameter s, there exists a unique positively oriented orthonormal frame $\{\gamma', N_\gamma\}$, called the Frenet frame of γ. The Frenet formulas of γ with respect to $\{\gamma', N_\gamma\}$ are

$$\begin{cases} \nabla_{\gamma'} \gamma' = K N_\gamma \\ \nabla_{\gamma'} N_\gamma = K \gamma', \end{cases} \tag{12}$$

where $K(s)$ is the curvature function of γ.

Lemma 3.1. *Let $\gamma(s)$ be a timelike curve on (M, J, g) which is parametrized by its arc length parameter s. We set a function ρ along γ by $\rho(s) = g(\gamma', J\gamma')$.*

(1) *With respect to the basis $\{\gamma', J\gamma'\}$ along γ, the Frenet frame $\{\gamma', N_\gamma\}$ and the curvature K of γ are determined as follows:*

(a) *If $\{\gamma', J\gamma'\}$ is positively (resp. negatively) oriented, then*

$$N_\gamma = \frac{1}{\sqrt{1+\rho^2}}(\rho\gamma' + J\gamma') \quad \left(resp.\ N_\gamma = -\frac{1}{\sqrt{1+\rho^2}}(\rho\gamma' + J\gamma')\right);$$

(b) *If $\{\gamma', J\gamma'\}$ is positively (resp. negatively) oriented, then*

$$K = \frac{1}{2\sqrt{1+\rho^2}}\frac{d\rho}{ds} \quad \left(resp.\ K = -\frac{1}{2\sqrt{1+\rho^2}}\frac{d\rho}{ds}\right). \tag{13}$$

(2) *$\gamma(s)$ is geodesic if and only if ρ is a constant.*

(3) *If $\{\gamma', J\gamma'\}$ is positively oriented, then $K = -1$ (resp. $K = 1$) if and only if $\rho = \rho_1$ (resp. $\rho = \rho_2$), where*

$$\rho_1 = \frac{A^4 - e^{4s}}{2A^2 e^{2s}}, \qquad \rho_2 = \frac{D^4 e^{4s} - 1}{2D^2 e^{2s}}, \qquad A > 0, D > 0. \tag{14}$$

If $\{\gamma', J\gamma'\}$ is negatively oriented, then $K = -1$ (resp. $K = 1$) if and only if $\rho = \rho_2$ (resp. $\rho = \rho_1$).

Proof. (1) By the conditions $g(N_\gamma, N_\gamma) = 1$ and $g(\gamma', N_\gamma) = 0$ we find

$$N_\gamma = \pm \frac{\rho}{\sqrt{1 + \rho^2}} \gamma' \pm \frac{1}{\sqrt{1 + \rho^2}} J\gamma'.$$

One can easily check the assertion (a). By the first equation in (12) we have $K = g(\nabla_{\gamma'}\gamma', N_\gamma)$. We hence obtain

$$K = \frac{1}{\sqrt{1 + \rho^2}} g(\nabla_{\gamma'}\gamma', J\gamma') \quad \left(\text{resp.} K = -\frac{1}{\sqrt{1 + \rho^2}} g(\nabla_{\gamma'}\gamma', J\gamma') \right). \tag{15}$$

Since $\nabla J = J\nabla$ we have

$$\frac{d\rho}{ds} = \gamma' \circ g(\gamma', J\gamma') = 2g(\nabla_{\gamma'}\gamma', J\gamma'). \tag{16}$$

These equalities (15) and (16) lead us to (13).

(2) The first equation in (12) shows that $\gamma(s)$ is geodesic if and only if $K = 0$, which is equivalent to the condition that ρ is a constant by (13).

(3) Since we have

$$\int \frac{d\rho}{\sqrt{1 + \rho^2}} = \ln(\rho + \sqrt{1 + \rho^2}) + c_0 = \sinh^{-1}\rho + c_0,$$

with a constant c_0, we get the assertion by use of (13). □

Further, we treat the case when $\{\gamma', J\gamma'\}$ is positively oriented. Hence N_γ and K are given as in Lemma 3.1 (1).

Let $\gamma(s)$ be a timelike curve on (M, J, g) parametrized by its arc length parameter s and $V(s)$ be a timelike unit vector field along γ. Then V is one of the four vector fields $\pm X, \pm Y$, which are expressed by use of the Frenet frame $\{\gamma', N_\gamma\}$ of γ as

$$X = \sqrt{1 + \mu^2}\gamma' + \mu N_\gamma, \tag{17}$$

$$Y = \sqrt{1 + \mu^2}\gamma' - \mu N_\gamma, \tag{18}$$

with an arbitrary smooth function $\mu = \mu(s)$ along γ. The vector field Y is obtained from X by replacing μ with $-\mu$ in (17). In the same way $-Y$

is obtained from $-X$. We therefore need to consider only the case when $V - \perp X$.

Theorem 3.1. *Let $\gamma(s)$ be a timelike curve on (M, J, g) parametrized by the arc length parameter s and $X(s)$ be a timelike unit vector field along γ determined by* (17). *The curves $C(s) = (\gamma(s), X(s))$ and $\overline{C}(s) = (\gamma(s), -X(s))$ on $(U(TM), g')$ are null curves of first type if and only if $\mu = \mu_1$ or $\mu = \mu_2$ which are given by*

$$\mu_1 = \frac{A_1^2 e^{2s} - B^2}{2A_1 Be^s}, \qquad \mu_2 = \frac{A_2^2 e^{-2s} - B^2}{2A_2 Be^{-s}}, \qquad (19)$$

where $B = \sqrt{\rho(s) + \sqrt{1 + \rho(s)^2}}$ and A_1, A_2 are arbitrary positive constants.

Proof. Since $\gamma(s)$ is a timelike curve on M and $g(\gamma', \gamma') = -1$, from Proposition 3.1 it follows that $C(s)$ and $\overline{C}(s)$ are null curves of first type if and only if the following condition holds

$$g(\nabla_{\gamma'} X, \nabla_{\gamma'} X) = 1. \qquad (20)$$

By using (12) and (17) we find

$$\nabla_{\gamma'} X = \Big(\frac{\mu}{\sqrt{1 + \mu^2}} \frac{d\mu}{ds} + \mu K \Big)\gamma' + \Big(K\sqrt{1 + \mu^2} + \frac{d\mu}{ds} \Big)N_\gamma. \qquad (21)$$

Substituting this into (20) we obtain

$$\Big(\frac{d\mu}{ds} \Big)^2 + 2K\sqrt{1 + \mu^2}\, \frac{d\mu}{ds} + (K^2 - 1)(1 + \mu^2) = 0.$$

Since it is decomposed into two ordinary differential equations

$$\frac{d\mu}{ds} = \sqrt{1 + \mu^2}(1 - K), \qquad \frac{d\mu}{ds} = -\sqrt{1 + \mu^2}(1 + K),$$

by substituting (13) into these differential equations, we obtain that their solutions are the functions μ_1 and μ_2, respectively. □

For a timelike curve $\gamma(s)$ on M, Theorem 3.1 tells us that the four curves $C_i(s) = (\gamma(s), X_i(s))$ and $\overline{C}_i(s) = (\gamma(s), -X_i(s))$ $(i = 1, 2)$ with

$$X_i = \sqrt{1 + \mu_i^2}\gamma' + \mu_i N_\gamma, \qquad i = 1, 2. \qquad (22)$$

are the only null curves of first type on $(U(TM), g')$ determined by γ and timelike unit vector fields. We here study the case when they are geodesics.

Proposition 3.2. *For the null curves of first type $C_i(s)$ and $\overline{C}_i(s)$ $(i = 1, 2)$ on $(U(TM), g')$ the following conditions are mutually equivalent:*

1) $C_1(s)$ *is geodesic*;
2) $C_2(s)$ *is geodesic*;
3) $\overline{C}_1(s)$ *is geodesic*;
4) $\overline{C}_2(s)$ *is geodesic*;
5) $\gamma(s)$ *is geodesic*;
6) *The function* $\rho = g(\gamma', J\gamma')$ *is a constant function.*

Proof. By using (21) and

$$\frac{d\mu_1}{ds} = \sqrt{1 + \mu_1^2}(1 - K), \qquad \frac{d\mu_2}{ds} = -\sqrt{1 + \mu_2^2}(1 + K) \qquad (23)$$

we get

$$\nabla_{\gamma'}(\pm X_1) = \pm\left(\mu_1\gamma' + \sqrt{1 + \mu_1^2}N_\gamma\right),$$

$$\nabla_{\gamma'}(\pm X_2) = \mp\left(\mu_2\gamma' + \sqrt{1 + \mu_2^2}N_\gamma\right),$$

respectively. By differentiating these equalities and by using (23), we find

$$\nabla_{\gamma'}\nabla_{\gamma'}(\pm X_i) = \pm X_i, \qquad i = 1, 2. \qquad (24)$$

Substituting (24) and $g(\nabla_{\gamma'}(\pm X_i), \nabla_{\gamma'}(\pm X_i)) = 1$ $(i = 1, 2)$ in (11) we obtain

$$\nabla'_{\dot{C}_i}\dot{C}_i = \nabla'_{\dot{\overline{C}}_i}\dot{\overline{C}}_i = (\nabla_{\gamma'}\gamma')^h, \qquad i = 1, 2. \qquad (25)$$

This shows that conditions 1) – 5) are mutually equivalent. Since Lemma 3.1 guarantees that 5) and 6) are equavalent, we complete the proof. □

Next, we endow the tangent bundle TM of (M, J, g) with an almost complex structure $\overline{J}$ and then $U(TM)$ with an almost contact B-metric structure. We consider the almost complex structure $\overline{J}$ on TM defined in [6] by

$$\overline{J}X^h = (JX)^v, \qquad \overline{J}X^v = (JX)^h. \qquad (26)$$

We directly verify that $\overline{g}^S$ satisfies $\overline{g}^S(\overline{J}X^h, \overline{J}Y^h) = -\overline{g}^S(X^h, Y^h)$, $\overline{g}^S(\overline{J}X^v, \overline{J}Y^v) = -\overline{g}^S(X^v, Y^v)$ and $\overline{g}^S(\overline{J}X^h, \overline{J}Y^v) = 0$, which means that $\overline{g}^S$ is a Norden metric on TM. Hence, $(TM, \overline{J}, \overline{g}^S)$ is a 4-dimensional almost complex manifold with Norden metric. Since the base manifold (M, J, g) is a flat Kähler-Norden manifold, we can easily check that $(TM, \overline{J}, \overline{g}^S)$ is a Kähler-Norden manifold. Now, we consider the unit tangent bundle $U(TM)$. In local coordinates the vector field $\overline{JN}$ is given by $\overline{u}^i\left(J\frac{\partial}{\partial x^i}\right)^h$.

Thus, we obtain $\overline{g}^S(N, \overline{JN}) = 0$, which shows that $\overline{JN}$ is a tangent vector field of $U(TM)$. We hence find that $U(TM)$ can be endowed with the following almost contact structure (φ, ξ, η) given by

$$\xi = -\overline{JN}, \quad \varphi X' = \overline{J}X' + \overline{g}^S(X', \overline{JN})N, \quad \eta(X') = -\overline{g}^S(X', \overline{JN}), \quad (27)$$

where X' is an arbitrary vector field on $U(TM)$ (see [5]). It is easy to check that the metric g' induced by the Norden metric $\overline{g}^S$ is a B-metric on $U(TM)$. Hence, the 3-dimensional Lorentzian manifold $(U(TM), g')$ is an almost contact B-metric manifold, which we denote by $(U(TM), \varphi, \xi, \eta, g')$. We here study the expressions of ξ and $\eta(\dot{C})$ at points on a curve $C(t) = (\gamma(t), X(t))$ on $(U(TM), \varphi, \xi, \eta, g')$. We have

$$\xi\big|_C = -(\overline{JN})\big|_C = -\overline{J}X^v = -(JX)^h. \quad (28)$$

By using (8), (27) and (28), we obtain

$$\eta(\dot{C}) = -g(JX, \dot{\gamma}) \circ \pi. \quad (29)$$

Theorem 3.2. *For the null curves of first type* $C_i(s) = (\gamma(s), X_i(s))$ *and* $\overline{C}_i(s) = (\gamma(s), -X_i(s))$ $(i = 1, 2)$ *on* $(U(TM), \varphi, \xi, \eta, g')$, *with the curvature function K of γ the following hold:*

(1) $\eta(\dot{C}_i) = -\eta(\dot{\overline{C}}_i)$ *for* $i = 1, 2$;
(2) $\eta(\dot{C}_1)$ *and* $\eta(\dot{\overline{C}}_1)$ *are constant functions if and only if* $K \equiv -1$;
(3) $\eta(\dot{C}_2)$ *and* $\eta(\dot{\overline{C}}_2)$ *are constant functions if and only if* $K \equiv 1$.

Proof. (1) Since

$$X_i = \sqrt{1 + \mu_i^2}\, \gamma' + \mu_i N_\gamma = \left(\sqrt{1 + \mu_i^2} + \frac{\mu_i \rho}{\sqrt{1 + \rho^2}}\right) \gamma' + \frac{\mu_i}{\sqrt{1 + \rho^2}} J\gamma'$$

by Lemma 3.1 (1) and by (22), by using (29) we have

$$\eta(\dot{C}_i) = -\eta(\dot{\overline{C}}_i) = -\rho\sqrt{1 + \mu_i^2} - \mu_i\sqrt{1 + \rho^2}, \quad i = 1, 2. \quad (30)$$

(2), (3) Differentiating both sides of (30) we obtain

$$\frac{d\eta(\dot{C}_i)}{ds} = -\frac{d\eta(\dot{\overline{C}}_i)}{ds} = -\left(\frac{\mu_i}{\sqrt{1 + \mu_i^2}}\rho + \sqrt{1 + \rho^2}\right)\frac{d\mu_i}{ds}$$
$$-\left(\frac{\rho}{\sqrt{1 + \rho^2}}\mu_i + \sqrt{1 + \mu_i^2}\right)\frac{d\rho}{ds}, \quad i = 1, 2. \quad (31)$$

Substituting (23) in (31) and using (13) we get

$$\frac{d\eta(\dot{C}_1)}{ds} = -\frac{d\eta(\dot{\overline{C}}_1)}{ds} = -(K + 1)\left(\mu_1 \rho + \sqrt{1 + \mu_1^2}\sqrt{1 + \rho^2}\right), \quad (32)$$

$$\frac{d\eta(\dot{C}_2)}{ds} = -\frac{d\eta(\dot{\overline{C}}_2)}{ds} = -(K-1)\left(\mu_2\rho + \sqrt{1+\mu_2^2}\sqrt{1+\rho^2}\right). \quad (33)$$

The equality (32) shows that $\dfrac{d\eta(\dot{C}_1)}{ds} = -\dfrac{d\eta(\dot{\overline{C}}_1)}{ds} = 0$ if and only if $K \equiv -1$,

and the equality (33) shows that $\dfrac{d\eta(\dot{C}_2)}{ds} = -\dfrac{d\eta(\dot{\overline{C}}_2)}{ds} = 0$ if and only if $K \equiv 1$. This completes the proof.   □

As an immediate consequence of Proposition 3.2 and Theorem 3.2 we obtain

Corollary 3.1. *For* $i = 1, 2$, *if the function* $\eta(\dot{C}_i)$ *of the null curve of first type* $C_i(s)$ $(i = 1, 2)$ *on* $(U(TM), \varphi, \xi, \eta, g')$ *is constant, then* $C_i(s)$ *and* $\overline{C}_i(s)$ *are not geodesics.*

An important class of curves on the almost contact B-metric manifold $(U(TM), \varphi, \xi, \eta, g')$ is the class of so called Legendre curves. A curve $C(t)$ is said to be a *Legendre curve* if $\eta(\dot{C}) = 0$. The condition that a curve to be a Legendre curve does not depend on parametrizations of this curve. This property does not hold for curves satisfying $\eta(\dot{C}) = a$ with a given constant a.

Proposition 3.3. *Let* $\gamma(s)$ *be a timelike curve on* M *whose curvature function satisfies* $K \equiv -1$ (*resp.* $K \equiv 1$). *Then the null curves of first type* $C_1(s)$ *and* $\overline{C}_1(s)$ (*resp.* $C_2(s)$ *and* $\overline{C}_2(s)$) *on* $(U(TM), \varphi, \xi, \eta, g')$ *are Legendre curves if and only if* $\mu_1 = -\rho_1$ (*resp.* $\mu_2 = -\rho_2$), *where* ρ_i $(i = 1, 2)$ *and* μ_i $(i = 1, 2)$ *are determined by* (14) *and* (19), *respectively.*

Proof. We study the case when $K \equiv -1$. It follows from Lemma 3.1 that the function ρ is equal to ρ_1. Substituting $\rho = \rho_1$ in (30) we have

$$\eta(\dot{C}_i) = -\eta(\dot{\overline{C}}_i) = -\rho_1\sqrt{1+\mu_i^2} - \mu_i\sqrt{1+\rho_1^2}, \quad i = 1, 2. \quad (34)$$

According to Theorem 3.2, $\eta(\dot{C}_1)$ and $\eta(\dot{\overline{C}}_1)$ are constants. Hence, $C_1(s)$ and $\overline{C}_1(s)$ are Legendre curves if and only if $\eta(\dot{C}_1) = \eta(\dot{\overline{C}}_1) = 0$. This condition is equivalent to $\mu_1 = -\rho_1$ by (34). We can show the case when $K \equiv 1$ analogously.   □

The following is a particular case of Corollary 3.1

Corollary 3.2. *Legendre null curves of first type* $C_i(s)$ *and* $\overline{C}_i(s)$ $(i = 1, 2)$ *on* $(U(TM), \varphi, \xi, \eta, g')$ *are not geodesics.*

4. Cartan framed null curves with respect to the original parameter on the unit tangent bundle with a unit timelike normal vector field of a two-dimensional Kähler-Norden manifold

It is known [3] that for a null curve $C(t)$ on a 3-dimensional Lorentzian manifold (L, g) a general Frenet frame $F = \{\dot{C}, N, W\}$ along C is determined by

$$g(\dot{C}, N) = g(W, W) = 1, \quad g(N, N) = g(N, W) = g(\dot{C}, W) = 0. \quad (35)$$

The general Frenet formulas with respect to F and the Levi-Civita connection ∇ of (L, g) are

$$\begin{cases} \nabla_{\dot{C}}\dot{C} = h\dot{C} + k_1 W, \\ \nabla_{\dot{C}}N = -hN + k_2 W, \\ \nabla_{\dot{C}}W = -k_2\dot{C} - k_1 N, \end{cases} \quad (36)$$

where h, k_1 and k_2 are smooth functions on L. The functions k_1 and k_2 are called *curvature functions* of C.

It is well known that the Frenet frame F and the Frenet formulas (36) with respect to F are not unique as they depend on the parameter and the choice of the screen vector bundle of C (for details see [2, pp. 56-58], [3, pp. 25-29]). In [2, p. 58] it is proved that there exists a parameter p for which the function h vanishes in (36). This parameter is called a *distinguished parameter* of C. The pair $(C(p), F(p))$, where F is a Frenet frame along C with respect to a distinguished parameter p, is called a *framed null curve* (see [3]). Since $(C(p), F(p))$ is not unique in general, we look for a Frenet frame of C for which (36) have minimum number of curvature functions, i.e.

$$\begin{cases} \nabla_{\dot{C}}\dot{C} = W, \\ \nabla_{\dot{C}}N = \tau W, \\ \nabla_{\dot{C}}W = -\tau\dot{C} - N. \end{cases} \quad (37)$$

Such a frame is called the *Cartan Frenet frame* of a non-geodesic null curve $C(p)$, which is unique by the assumption $g(\ddot{C}, \ddot{C}) = k_1 = 1$. The corresponding equations (37) are called the *Cartan Frenet formulas* of $C(p)$ and τ is called a *torsion function*. A null curve together with its Cartan Frenet frame is called a *Cartan framed null curve*.

Proposition 4.1. *Let* $C_i(s) = (\gamma(s), X_i(s))$ *and* $\overline{C}_i(s) = (\gamma(s), -X_i(s))$ $(i = 1, 2)$ *be non-geodesic null curves of first type on* $(U(TM), g')$. *Then* *there exist Frenet frames* $F_i = \{\dot{C}_i, N_i, W_i\}$ *and* $\overline{F}_i = \{\dot{\overline{C}}_i, \overline{N}_i, \overline{W}_i\}$ *(i =* $1, 2$*) along* $C_i(s)$ *and* $\overline{C}_i(s)$*, respectively, which are determined by*

$$W_i = (N_\gamma)^h, \quad N_i = \frac{1}{2}\left(-(\gamma')^h + (\nabla_{\gamma'} X_i)^v\right), \quad i = 1, 2, \qquad (38)$$

$$\overline{W}_i = (N_\gamma)^h, \quad \overline{N}_i = -\frac{1}{2}\left((\gamma')^h + (\nabla_{\gamma'} X_i)^v\right), \quad i = 1, 2, \qquad (39)$$

such that the pairs $(C_i(s), F_i(s))$ *and* $\left(\overline{C}_i(s), \overline{F}_i(s)\right)$ *are framed null curves with respect to the original parameter* s*. If we denote by* $(k_1)_i$*,* $(k_2)_i$ *and* $(\overline{k}_1)_i$*,* $(\overline{k}_2)_i$ *(i = 1, 2) the curvature functions of* $C_i(s)$ *and* $\overline{C}_i(s)$ *with respect to* $F_i(s)$ *and* $\overline{F}_i(s)$*, respectively, then they satisfy* $(k_1)_i = (\overline{k}_1)_i = K$*,* $(k_2)_i = (\overline{k}_2)_i = -K/2$*.*

Proof. It follows from Proposition 3.2 that γ is non-geodesic. By using the first formula in (12) the equality (25) becomes

$$\nabla'_{\dot{C}_i} \dot{C}_i = \nabla'_{\dot{\overline{C}}_i} \dot{\overline{C}}_i = K(N_\gamma)^h, \qquad i = 1, 2. \qquad (40)$$

Comparing (40) with the first formula in (36) and taking into account that $\overline{g}^S((N_\gamma)^h, (N_\gamma)^h) = 1$, we may choose $W_i = \overline{W}_i = (N_\gamma)^h$ $(i = 1, 2)$. Then we get $h_i = \overline{h}_i = 0$ and $(k_1)_i = (\overline{k}_1)_i = K$, $(i = 1, 2)$. The vanishing of h_i (resp. $\overline{h}_i$) means that the original parameter s is a distinguished parameter with respect to the Frenet frame F_i (resp. $\overline{F}_i$) $(i = 1, 2)$, where F_i (resp. $\overline{F}_i$) consists of $\dot{C}_i, W_i$ (resp. $\dot{\overline{C}}_i, \overline{W}_i$) and the vector field N_i (resp. $\overline{N}_i$) corresponding to W_i (resp. $\overline{W}_i$).

We now study the expressions of N_i and $\overline{N}_i$ $(i = 1, 2)$. First, replacing X in (8) with $(-X)$ we have

$$\dot{\overline{C}}_i = \left((\gamma')^h - (\nabla_{\gamma'} X_i)^v\right)(\overline{C}_i(s)), \quad i = 1, 2. \qquad (41)$$

Now, using (5), (7),(8), (12) and (41) we obtain

$$\nabla'_{\dot{C}_i} W_i = \nabla'_{\dot{\overline{C}}_i} \overline{W}_i = K(\gamma')^h, \quad i = 1, 2. \qquad (42)$$

On the other hand, according to the third formula in (36), we have

$$\nabla'_{\dot{C}_i} W_i = -(k_2)_i \dot{C}_i - KN_i, \quad i = 1, 2, \qquad (43)$$

$$\nabla'_{\dot{\overline{C}}_i} \overline{W}_i = -(\overline{k}_2)_i \dot{\overline{C}}_i - K\overline{N}_i, \quad i = 1, 2. \qquad (44)$$

From (43) and (44) we derive

$$(k_2)_i = \frac{1}{2K} g(\nabla'_{\dot{C}_i} W_i, \nabla'_{\dot{C}_i} W_i), \quad i = 1, 2,$$

$$(\overline{k}_2)_i = \frac{1}{2K} g(\nabla'_{\dot{\overline{C}}_i} \overline{W}_i, \nabla'_{\dot{\overline{C}}_i} \overline{W}_i), \quad i = 1, 2.$$

By using (42) we get

$$g(\nabla'_{\dot{C}_i} W_i, \nabla'_{\dot{C}_i} W_i) = g(\nabla'_{\dot{\overline{C}}_i} \overline{W}_i, \nabla'_{\dot{\overline{C}}_i} \overline{W}_i) = -K^2 \quad (i = 1, 2).$$

We hence obtain $(k_2)_i = (\overline{k}_2)_i = -K/2$. Finally, taking into account (8), (41) and (42) we obtain the expressions of N_i and $\overline{N}_i$ $(i = 1, 2)$. □

According to Proposition 4.1, the Frenet formulas of non-geodesic null curves of first type $C_i(s)$ and $\overline{C}_i(s)$ $(i = 1, 2)$ with respect to the Frenet frames $F_i = \{\dot{C}_i, N_i, W_i\}$ and $\overline{F}_i = \{\dot{\overline{C}}_i, \overline{N}_i, \overline{W}_i\}$, determined by (38) and (39) are

$$\begin{cases} \nabla'_{\dot{C}_i} \dot{C}_i = K W_i, \\[2mm] \nabla'_{\dot{C}_i} N_i = -\dfrac{K}{2} W_i, \\[2mm] \nabla'_{\dot{C}_i} W_i = \dfrac{K}{2} \dot{C}_i - K N_i, \end{cases} \qquad \begin{cases} \nabla'_{\dot{\overline{C}}_i} \dot{\overline{C}}_i = K \overline{W}_i, \\[2mm] \nabla'_{\dot{\overline{C}}_i} \overline{N}_i = -\dfrac{K}{2} \overline{W}_i, \\[2mm] \nabla'_{\dot{\overline{C}}_i} \overline{W}_i = \dfrac{K}{2} \dot{\overline{C}}_i - K \overline{N}_i, \end{cases}$$

respectively. An immediate consequence from Theorem 3.2 and the above formulas we have

Proposition 4.2. *Let $C_i(s)$ and $\overline{C}_i(s)$ $(i = 1, 2)$ be null curves of first type on $(U(TM), \varphi, \xi, \eta, g')$ for which $\eta(\dot{C}_i)$ $(i = 1, 2)$ are constants. Then $\{C_1(s), F'_1(s)\}, \{\overline{C}_1(s), \overline{F}'_1(s)\}$ and $\{C_2(s), F_2(s)\}, \{\overline{C}_2(s), \overline{F}_2(s)\}$ are Cartan framed null curves with respect to the original parameter s. The Cartan Frenet frames of $C_1, \overline{C}_1$ and $C_2, \overline{C}_2$ are*

$$F'_1 = \{\dot{C}_1, -W_1, N_1\}, \qquad \overline{F}'_1 = \{\dot{\overline{C}}_1, -\overline{W}_1, \overline{N}_1\},$$
$$F_2 = \{\dot{C}_2, W_2, N_2\}, \qquad \overline{F}_2 = \{\dot{\overline{C}}_2, \overline{W}_2, \overline{N}_2\},$$

respectively, where W_i, N_i $(i = 1, 2)$ and $\overline{W}_i, \overline{N}_i$ $(i = 1, 2)$ are determined by (38) and (39). The torsion of $C_i(s)$ and $\overline{C}_i(s)$ $(i = 1, 2)$ is $\tau_i = \overline{\tau}_i = -1/2$.

Corollary 4.1. *Legendre null curves of first type $C_i(s)$ and $\overline{C}_i(s)$ $(i = 1, 2)$ on $(U(TM), \varphi, \xi, \eta, g')$ are Cartan framed null curves with respect to the original parameter s.*

5. Examples of null curves and Legendre null curves of first type on $(U(T\mathbb{R}^2), \varphi, \xi, \eta, g')$

We consider $(\mathbb{R}^2, J, g)$, where the almost complex structure J and the Norden metric g are defined with respect to the local basis $\left\{\dfrac{\partial}{\partial x_1}, \dfrac{\partial}{\partial x_2}\right\}$ on

$U\left(\subset \mathbb{R}^2\right)$, as follows:

$$J\left(\frac{\partial}{\partial x_1}\right) = \frac{\partial}{\partial x_2}, \quad J\left(\frac{\partial}{\partial x_2}\right) = -\frac{\partial}{\partial x_1},$$

$$g\left(\frac{\partial}{\partial x_1}, \frac{\partial}{\partial x_1}\right) = -g\left(\frac{\partial}{\partial x_2}, \frac{\partial}{\partial x_2}\right) = -1, \quad g\left(\frac{\partial}{\partial x_1}, \frac{\partial}{\partial x_2}\right) = 0. \tag{45}$$

We directly check that the 2-dimensional almost complex manifold with Norden metric $(\mathbb{R}^2, J, g)$ is Kähler-Norden. Taking into account that $\mathbb{R}^2$ is flat, from (4) we obtain

$$\left(\frac{\partial}{\partial x^i}\right)^h = \frac{\partial}{\partial \overline{x}^i}, \quad \left(\frac{\partial}{\partial x^i}\right)^v = \frac{\partial}{\partial \overline{u}^i}, \quad i = 1, 2. \tag{46}$$

By using (26), (45) and (46), for $\overline{J}$ on $\pi^{-1}(U)$ with respect to the local basis $\left\{\frac{\partial}{\partial \overline{x}^1}, \frac{\partial}{\partial \overline{x}^2}, \frac{\partial}{\partial \overline{u}^1}, \frac{\partial}{\partial \overline{u}^2}\right\}$ we have

$$\overline{J}\left(\frac{\partial}{\partial \overline{x}^1}\right) = \frac{\partial}{\partial \overline{u}^2}, \quad \overline{J}\left(\frac{\partial}{\partial \overline{x}^2}\right) = -\frac{\partial}{\partial \overline{u}^1}, \quad \overline{J}\left(\frac{\partial}{\partial \overline{u}^1}\right) = \frac{\partial}{\partial \overline{x}^2}, \quad \overline{J}\left(\frac{\partial}{\partial \overline{u}^2}\right) = -\frac{\partial}{\partial \overline{x}^1}.$$

The unit tangent bundle $U(T\mathbb{R}^2)$ of the 4-dimensional Kähler-Norden manifold $(T\mathbb{R}^2, \overline{J}, \overline{g}^S)$ is defined by (6). We endow $U(T\mathbb{R}^2)$ with the almost contact B-metric structure (φ, ξ, η, g'), determined by (27).

We consider the timelike curve $\gamma = (\sinh s, \cosh s)$ on $(\mathbb{R}^2, J, g)$, parametrized by the arc length parameter. The vectors from the Frenet frame $\{\gamma', N_\gamma\}$ and the curvature K of γ are

$$\gamma' = (\cosh s, \sinh s), \qquad N_\gamma = (\sinh s, \cosh s), \qquad K = 1. \tag{47}$$

By using (45) we find $J\gamma' = (-\sinh s, \cosh s)$. Hence $\rho = \sinh 2s$. Then the functions μ_1 and μ_2 determined by (19) are

$$\mu_1 = \frac{A_1^2 - 1}{2A_1}, \quad A_1 > 0; \qquad \mu_2 = \frac{A_2^2 e^{-2s} - e^{2s}}{2A_2}, \quad A_2 > 0. \tag{48}$$

Substituting (48) in (22) we get

$$X_1 = \left(\frac{A_1^2 e^{2s} + 1}{2A_1 e^s}, \frac{A_1^2 e^{2s} - 1}{2A_1 e^s}\right), \quad X_2 = \left(\frac{A_2^2 e^{-s} + e^s}{2A_2}, \frac{A_2^2 e^{-s} - e^s}{2A_2}\right). \tag{49}$$

Now, it follows from Theorem 3.1 that the curves $C_i(s) = (\gamma(s), X_i(s))$ and $\overline{C}_i(s) = (\gamma(s), -X_i(s))$ $(i = 1, 2)$, where X_i are determined by (49), are null curves of first type on $(U(T\mathbb{R}^2), \varphi, \xi, \eta, g')$. Since $K = 1$, it follows from Theorem 3.2 that $\eta(\dot{C}_2)$ and $\eta(\dot{\overline{C}}_2)$ are opposite constants. By direct computations we obtain $\eta(\dot{C}_2) = -\eta(\dot{\overline{C}}_2) = (1 - A_2^2)/(2A_2)$. Moreover, according to Proposition 3.3, $C_2(s) = (\gamma(s), X_2(s))$ and $\overline{C}_2(s) = (\gamma(s), -X_2(s))$

are Legendre null curves if $\mu_2 = -\rho = -\sinh 2s$. We substitute $(-\sinh 2s)$ into (22) for μ_2 and obtain the following vector field

$$Y_2 = (\cosh s, -\sinh s).$$

The curves $\Gamma_2(s) = (\gamma(s), Y_2(s))$ and $\overline{\Gamma}_2(s) = (\gamma(s), -Y_2(s))$ are Legendre null curves of first type on $(U(T\mathbb{R}^2), \varphi, \xi, \eta, g')$. These curves are Cartan framed null curves with respect to the original parameter s. By using Proposition 4.2, (38) and (39) we find that the Cartan Frenet frames $F_2 = \{\dot{\Gamma}_2, W_2, N_2\}$ and $\overline{F}_2 = \{\dot{\overline{\Gamma}}_2, \overline{W}_2, \overline{N}_2\}$ of $\Gamma_2(s)$ and $\overline{\Gamma}_2(s)$, respectively, are expressed with respect to the local basis $\left\{ \dfrac{\partial}{\partial \overline{x}^1}, \dfrac{\partial}{\partial \overline{x}^2}, \dfrac{\partial}{\partial \overline{u}^1}, \dfrac{\partial}{\partial \overline{u}^2} \right\}$ as follows:

$$\dot{\Gamma}_2 = (\cosh s, \sinh s, \sinh s, -\cosh s), \quad W_2 = (\sinh s, \cosh s, 0, 0),$$

$$N_2 = \left(-\frac{1}{2}\cosh s, -\frac{1}{2}\sinh s, \frac{1}{2}\sinh s, -\frac{1}{2}\cosh s \right);$$

$$\dot{\overline{\Gamma}}_2 = (\cosh s, \sinh s, -\sinh s, \cosh s), \quad \overline{W}_2 = W_2,$$

$$\overline{N}_2 = \left(-\frac{1}{2}\cosh s, -\frac{1}{2}\sinh s, -\frac{1}{2}\sinh s, \frac{1}{2}\cosh s \right).$$

References

[1] E. Boeckx & L. Vanhecke, Characteristic reflections on unit tangent sphere bundles, *Houston J. Math.*, **23**, no. 3, 427–448, (1997).

[2] K. L. Duggal & A. Bejancu, *Lightlike Submanifolds of Semi-Riemannian Manifolds and Applications*, Kluwer Academic, (1996).

[3] K.L. Duggal & D.H. Jin, *Null Curves and Hypersurfaces of Semi-Riemannian Manifolds*, World Scientific, Singapore, (2007).

[4] G. Ganchev, V. Mihova & K. Gribachev, Almost contact manifolds with B-metric, *Math. Balkanica*, **7**, 262–276, (1993).

[5] M. Manev, Almost contact B-metric hypersurfaces of Kählerian manifolds with B-metric, *Perspectives of Complex analysis, Differential Geometry and Mathematical Physics*, S. Dimiev & K. Sekigawa eds., 159–170, World Scientific, Singapore, 2001.

[6] M. Manev, Tangent bundles with Sasaki metric and almost hypercomplex pseudo-Hermitian structure, *Topics in Almost Hermitian geometry and related fields*, Y. Matsushita, E. G. Rio, H. Hashimoto, T. Koda & T. Oguro eds., 170–185, World Scientific, Singapore, 2004.

[7] K. Yano & S. Ishihara, *Tangent and Cotangent Bundles*, Marcel Dekker, New York, (1973).

Received December 19, 2016
Revised May 19, 2017

Contemporary Perspectives
in Differential Geometry
and its Related Fields   131 – 149

HORIZONTAL LIFTS OF CURVES
THROUGH A HOPF FIBRATION
AND SOME EXAMPLES OF HOPF TORI

Yoshitaka FUKADA

*Division of Mathematics and Mathematical Science,
Nagoya Institute of Technology,
Nagoya 466-8555, Japan
E-mail: e0929080_ccalumni@yahoo.co.jp*

We study Hopf tori obtained from a Viviani curve, small circles and a great circle in S^2. By using expressions of horizontal lifts of these curves we give parametrizations of these Hopf tori, and show some geometric properties.

Keywords: Hopf fibration; Horizontal curves; Riemmanian submersion; Hopf torus.

1. Introduction

Let $\pi : S^3 \to S^2$ be a Hopf fibration of a 3-dimensional unit sphere to a 2-dimensional unit sphere. The main purposes of this paper is to give representations of the horizontal lifts of curves in S^2 with respect to a Hopf fibration, By using these horizontal lifts, we obtain flat surfaces which are called Hopf tori when they are compact. Especially, we give an explicit parametrization of the Hopf torus corresponding to a Viviani curve. By using this, we write down concretely the deformation of metrics from the induced one to the canonical Euclidean one.

The author would like to express his hearty thanks to Professors Hideya Hashimoto and Misa Ohashi for their valuable advice. Also, he is grateful to Professor Toshiaki Adachi for his encouragement.

2. Hopf fibration by quaternions

Let $\mathbb{H}$ be a skew field of quaternions which is isomorphic to a real 4-dimensional vector space with canonical inner product $\langle\ ,\ \rangle$. We fix an orthonormal basis $\{1, i, j, k\}$, It satisfies the following multiplication;

$$i^2 = j^2 = k^2 = -1,\ ij = -ji = k,\ jk = -kj = i,\ ki = -ik = j,$$

where 1 is an identity element of $\mathbb{H}$. The conjugation of $q \in \mathbb{H}$ is defined by $\bar{q} = 2\langle q, 1\rangle 1 - q$. The set $\mathbb{H}$ is a non-commutative, associative normed division algebra. We have $\langle qp, qp \rangle = \langle q, q \rangle \langle p, p \rangle$ for arbitrary $q, p \in \mathbb{H}$. For each $q \in S^3 = \{ q \in \mathbb{H} \mid \|q\| = 1 \}$, we have $qi\bar{q} \in \operatorname{Im} \mathbb{H} = \operatorname{span}_{\mathbb{R}} \{i,\ j,\ k\}$ and $\|qi\bar{q}\| = 1$, where $\| \cdot \|$ is the norm of $\mathbb{H}$ induced by the inner product $\langle\ ,\ \rangle$. We can therefore define a mapping $\pi : S^3 \to S^2$ as

$$\pi(q) = qi\bar{q}.$$

We note that this map satisfies

$$\pi(qe^{i\theta}) = qe^{i\theta}\, i\, \overline{qe^{i\theta}} = qi\bar{q} = \pi(q)$$

for all $q \in S^3$, $e^{i\theta} \in S^1$. Therefore S^3 is a S^1-bundle over S^2. We call this map $\pi : S^3 \to S^2$ a *Hopf fibration*. The purpose of this paper is to give an explicit expressions of horizontal lifts in S^3 for regular curves in S^2. In order to define horizontal lifts of curves through the Hopf map π, we recall some fundamental properties of the tangent space of a 3-dimensional sphere S^3. Since S^3 is included in $\mathbb{H}$, we have

$$T_q S^3 = \operatorname{span}_{\mathbb{R}} \{qi,\ qj,\ qk\}$$

at each $q \in S^3$. Note that $\{qi,\ qj,\ qk\}$ is an orthonormal frame at $q \in S^3$. The Hopf fibration $\pi : S^3 \to S^2$ admits a structure of Riemannian submersion in the following way. First, we give vertical and horizontal distributions $\mathcal{V}$ and $\mathcal{H}$ on S^3 as follows. Since $\left.\dfrac{d}{d\theta}\right|_{\theta=0}(qe^{i\theta}) = qi$, the vertical distribution is given by $\mathcal{V}_q = \operatorname{span}_{\mathbb{R}}\{qi\}$. Therefore we define the horizontal distribution as $\mathcal{H}_q = \operatorname{span}_{\mathbb{R}}\{qj,\ qk\}$. Clearly, we have $\mathcal{V}_q \perp \mathcal{H}_q$ and $T_q S^3 = \mathcal{H}_q \oplus \mathcal{V}_q$ at each $q \in S^3$. The horizontal distribution $\mathcal{H}_q$ satisfies $R_{e^{i\theta}*} \mathcal{H}_q = \mathcal{H}_{R_{e^{-i\theta}} q}$, where $R_a : \mathbb{H} \to \mathbb{H}$ denotes a right multiplication Also, we note that

$$\pi_*(qj) = -2qk\bar{q}, \qquad \pi_*(qk) = 2qj\bar{q}.$$

Therefore the canonical metrics $\langle\ ,\ \rangle_{S^3}$ on S^3 and $\langle\ ,\ \rangle_{S^2}$ on S^2 are related by

$$\langle\ ,\ \rangle_{S^3} = \frac{1}{4}\pi^*\langle\ ,\ \rangle_{S^2} + \nu \odot \nu$$

where ν is a 1-form defined by $\nu(X) = \langle qi, X \rangle$ for arbitrary $X \in T_q S^3$ at an arbitrary point $q \in S^3$. Hence π induce the Riemannian submersion.

Next we give horizontal lifts of vector fields on S^2 and those of curves in S^2. For an arbitrary curve $\gamma : \mathbb{R} \to S^2$ $(\subset \operatorname{Im} \mathbb{H})$ in S^2, its inverse image

$$\pi^{-1}(\gamma(\mathbb{R})) = \{ q \in S^3 \mid \pi(q) \in \gamma(\mathbb{R}) \}$$

is called the *Hopf cylinder* for γ. It is diffeomorphic to the product manifold $\gamma(\mathbb{R}) \times S^1$. When a curve in S^2 is closed, then its inverse image is compact. We hence call it a *Hopf torus* in this case. Given $q \in S^3$, if ther exists a smooth curve $\eta : \mathbb{R} \to \pi^{-1}(\gamma(S^1))$ $(\subset S^3)$ satisfying

$$\pi \circ \eta = \gamma, \ \eta(0) = q \quad \text{and} \quad \left\langle \eta_*\left(\frac{d}{dt}\right), \eta(t)i \right\rangle \equiv 0 \quad \text{for all } t \in \mathbb{R},$$

then we call it a *horizontal lift* of γ passing through q. Next we study horizontal lifts of a vector field on S^2 and their representation related to the spherical coordinates of S^2. Let S^2 be the unit sphere in $(\mathbb{R}^3, (x, y, z))$, where (x, y, z) is the canonical orthogonal coordinate of $\mathbb{R}^3$. We set a local chart $\Phi : (-\pi/2, \pi/2) \times (-\pi, \pi) \to S^2$ as

$$\left(-\frac{\pi}{2}, \frac{\pi}{2}\right) \times (-\pi, \pi) \ni (\theta, \varphi) \mapsto \Phi(\theta, \varphi) = \begin{pmatrix} \cos\theta\cos\varphi \\ \cos\theta\sin\varphi \\ \sin\theta \end{pmatrix} \in S^2.$$

The induced metric on $(-\pi/2, \pi/2) \times (-\pi, \pi)$ is given by

$$ds^2 = d\theta^2 + \cos^2\theta \, d\varphi^2.$$

The local orthonormal frame field on S^2 is given by $\left\{ \frac{\partial}{\partial\theta}, \frac{1}{\cos\theta}\frac{\partial}{\partial\varphi} \right\}$. We compute the horizontal local vector field $\left(\Phi_*\frac{\partial}{\partial\theta}\right)^h$ on S^3 corresponding to the vector field $\frac{\partial}{\partial\theta}$. That is, it is a vector field satisfying

$$\pi_*\left(\left(\Phi_*\frac{\partial}{\partial\theta}\right)^h\right) = \Phi_*\left(\frac{\partial}{\partial\theta}\right) \quad \text{and} \quad \left\langle \left(\left(\Phi_*\frac{\partial}{\partial\theta}\right)^h\right)_q, qi \right\rangle = 0$$

at each $q \in S^3$. To do this, for given $\Phi(\theta, \varphi) \in S^2$, we solve the equation

$$\pi(q) = qi\bar{q} = \Phi(\theta, \varphi)$$

on q. Identifying $\mathbb{R}^3$ with $\text{Im}\mathbb{H}$, we have

$$\Phi(\theta, \varphi) \simeq \sin\theta i + j(\cos\theta e^{i\varphi}).$$

If we put $q = z_0 + jz_1$ with $z_0, z_1 \in \mathbb{C}$, we then have

$$\pi(q) = qi\bar{q} = i(|z_0|^2 - |z_1|^2) + j(2i\overline{z_0}z_1).$$

Hence the equation $\pi(q) = \Phi(\theta, \varphi)$ turns to the system of equations

$$|z_0|^2 + |z_1|^2 = 1, \ |z_0|^2 - |z_1|^2 = \sin\theta, \ 2i\overline{z_0}z_1 = \cos\theta e^{i\varphi}.$$

Thus, we find that the solution is

$$q = q(\theta, \varphi, \mu_0) = \left\{ \sin\left(\frac{\theta}{2} + \frac{\pi}{4}\right) + j\cos\left(\frac{\theta}{2} + \frac{\pi}{4}\right)(-i)e^{i\varphi} \right\} e^{i\mu_0} \quad (1)$$

with some fixed $\mu_0 \in \mathbb{R}$. We rewrite (1) in the following real form

$$
q = q(\theta, \varphi, \mu_0) = \begin{pmatrix} 1 & i & j & k \end{pmatrix} \begin{pmatrix} \sin\left(\dfrac{\theta}{2} + \dfrac{\pi}{4}\right) \cos\mu_0 \\[2mm] \sin\left(\dfrac{\theta}{2} + \dfrac{\pi}{4}\right) \sin\mu_0 \\[2mm] \cos\left(\dfrac{\theta}{2} + \dfrac{\pi}{4}\right) \sin(\varphi + \mu_0) \\[2mm] \cos\left(\dfrac{\theta}{2} + \dfrac{\pi}{4}\right) \cos(\varphi + \mu_0) \end{pmatrix}.
$$

This represents the inverse image of S^2. From this, we obtain the natural lift of a given curve of S^2. The local tangent vector field $q_* \dfrac{\partial}{\partial \theta}$ on S^3 is given by

$$
q_* \frac{\partial}{\partial \theta} = \begin{pmatrix} 1 & i & j & k \end{pmatrix} \frac{1}{2} \begin{pmatrix} \cos\left(\dfrac{\theta}{2} + \dfrac{\pi}{4}\right) \cos\mu_0 \\[2mm] \cos\left(\dfrac{\theta}{2} + \dfrac{\pi}{4}\right) \sin\mu_0 \\[2mm] -\sin\left(\dfrac{\theta}{2} + \dfrac{\pi}{4}\right) \sin(\varphi + \mu_0) \\[2mm] -\sin\left(\dfrac{\theta}{2} + \dfrac{\pi}{4}\right) \cos(\varphi + \mu_0) \end{pmatrix}. \tag{2}
$$

We shall show that this vector field coincides with $\left(\Phi_* \dfrac{\partial}{\partial \theta}\right)^h$ by direct calculation. Since

$$
q(\theta, \varphi, \mu_0)i = \begin{pmatrix} 1 & i & j & k \end{pmatrix} \begin{pmatrix} -\sin\left(\dfrac{\theta}{2} + \dfrac{\pi}{4}\right) \sin\mu_0 \\[2mm] \sin\left(\dfrac{\theta}{2} + \dfrac{\pi}{4}\right) \cos\mu_0 \\[2mm] \cos\left(\dfrac{\theta}{2} + \dfrac{\pi}{4}\right) \cos(\varphi + \mu_0) \\[2mm] -\cos\left(\dfrac{\theta}{2} + \dfrac{\pi}{4}\right) \sin(\varphi + \mu_0) \end{pmatrix}, \tag{3}
$$

we obtain

$$
\left\langle q_* \frac{\partial}{\partial \theta}, \, q(\theta, \varphi, \mu_0)i \right\rangle = 0 \tag{4}
$$

by (2). As we take $q(\theta, \varphi, \mu_0)$ so that $\pi\big(q(\theta, \varphi, \mu_0)\big) = \Phi(\theta, \varphi)$, we have $\pi_*\big(q_* \dfrac{\partial}{\partial \theta}\big) = \Phi_* \dfrac{\partial}{\partial \theta}$, hence find $\left(\Phi_* \dfrac{\partial}{\partial \theta}\right)^h = q_* \dfrac{\partial}{\partial \theta}$.

Since the curve $\Phi(\theta, \varphi_0)$ is a geodesic on S^2 for a fixed $\varphi_0 \in \mathbb{R}$, its horizontal lift is also a geodesic on S^3. Let $q_* \dfrac{\partial}{\partial \varphi}$ be the local tangent vector field on $\Phi\big((-\pi/2,\pi/2)\times(-\pi,\pi)\big)$ given by

$$
q_*\frac{\partial}{\partial\varphi} = \begin{pmatrix} 1 & i & j & k \end{pmatrix}\begin{pmatrix} 0 \\ 0 \\ \cos\Big(\dfrac{\theta}{2}+\dfrac{\pi}{4}\Big)\cos(\varphi+\mu_0) \\ -\cos\Big(\dfrac{\theta}{2}+\dfrac{\pi}{4}\Big)\sin(\varphi+\mu_0) \end{pmatrix}. \tag{5}
$$

By this expression and by (3), we have

$$
\Big\langle q_*\frac{\partial}{\partial\varphi},\; q(\theta,\varphi,\mu_0)i \Big\rangle = \frac{1}{2}(1-\sin\theta) = \cos^2\Big(\frac{\theta}{2}+\frac{\pi}{4}\Big). \tag{6}
$$

The horizontal lift of the local vector field $\Phi_*\dfrac{\partial}{\partial\varphi}$ of S^2 is hence given by

$$
\Big(\Phi_*\frac{\partial}{\partial\varphi}\Big)^h = q_*\frac{\partial}{\partial\varphi} - \Big\langle q_*\frac{\partial}{\partial\varphi},\; q(\theta,\varphi,\mu_0)i \Big\rangle q(\theta,\varphi,\mu_0)i
$$

$$
= \begin{pmatrix} 1 & i & j & k \end{pmatrix}\frac{\cos\theta}{2}\begin{pmatrix} \cos\Big(\dfrac{\theta}{2}+\dfrac{\pi}{4}\Big)\sin\mu_0 \\ -\cos\Big(\dfrac{\theta}{2}+\dfrac{\pi}{4}\Big)\cos\mu_0 \\ \sin\Big(\dfrac{\theta}{2}+\dfrac{\pi}{4}\Big)\cos(\varphi+\mu_0) \\ -\sin\Big(\dfrac{\theta}{2}+\dfrac{\pi}{4}\Big)\sin(\varphi+\mu_0) \end{pmatrix}.
$$

Thus, we obtain the following:

Proposition 2.1. *Let* $\gamma : I \to S^2$ *be an arbitrary* C^∞-*curve defined on an interval* $I \subset \mathbb{R}$. *If we express* $\gamma(t) = \Phi(\theta(t),\varphi(t))$ *for* $t \in I$, *then the horizontal lift* $\Big(\gamma_*\dfrac{d}{dt}\Big|_{t=0}\Big)^h$ *of the tangent vector* $\gamma_*\dfrac{d}{dt}\Big|_{t=0}$ *along* γ *is given as*

$$
\begin{pmatrix} 1 & i & j & k \end{pmatrix}\frac{1}{2}\begin{pmatrix} \cos\Big(\dfrac{\theta}{2}+\dfrac{\pi}{4}\Big)\Big(\cos\mu_0\dfrac{d\theta}{dt}\Big|_{t=0} + \cos\theta\sin\mu_0\dfrac{d\varphi}{dt}\Big|_{t=0}\Big) \\ \cos\Big(\dfrac{\theta}{2}+\dfrac{\pi}{4}\Big)\Big(\sin\mu_0\dfrac{d\theta}{dt}\Big|_{t=0} - \cos\theta\cos\mu_0\dfrac{d\varphi}{dt}\Big|_{t=0}\Big) \\ -\sin\Big(\dfrac{\theta}{2}+\dfrac{\pi}{4}\Big)\Big(\sin(\varphi+\mu_0)\dfrac{d\theta}{dt}\Big|_{t=0} - \cos\theta\cos(\varphi+\mu_0)\dfrac{d\varphi}{dt}\Big|_{t=0}\Big) \\ -\sin\Big(\dfrac{\theta}{2}+\dfrac{\pi}{4}\Big)\Big(\cos(\varphi+\mu_0)\dfrac{d\theta}{dt}\Big|_{t=0} + \cos\theta\sin(\varphi+\mu_0)\dfrac{d\varphi}{dt}\Big|_{t=0}\Big) \end{pmatrix}.
$$

Proof. As we have

$$
\gamma_*\frac{d}{dt}\Big|_{t=0} = \frac{d\theta}{dt}\Big|_{t=0}\Big(\Phi_*\frac{\partial}{\partial\theta}\Big) + \frac{d\varphi}{dt}\Big|_{t=0}\Big(\Phi_*\frac{\partial}{\partial\varphi}\Big),
$$

we get the asseertion by (2) and (5).   □

Next we give expressions of horizontal lifts of curves in S^2.

Proposition 2.2. *Let* $\gamma : I \to S^2$ *be an arbitrary* C^∞*-curve. If we express* $\gamma(t) = \Phi\big(\theta(t), \varphi(t)\big)$ *for* $t \in I$, *then its horizontal lift is given by*

$$\eta(t) = q(\theta(t),\ \varphi(t),\ \mu(t))$$
$$= \left\{ \sin\left(\frac{\theta(t)}{2} + \frac{\pi}{4}\right) + j\cos\left(\frac{\theta(t)}{2} + \frac{\pi}{4}\right)(-i)e^{i\varphi(t)} \right\} e^{i\mu(t)},$$

where $\mu(t)$ *is a solution of the following differential equation*

$$\frac{d\mu(t)}{dt} + \frac{1}{2}\big(1 - \sin\theta(t)\big)\frac{d\varphi(t)}{dt} = 0. \tag{7}$$

Proof. By (6) and by the expression of η, we obtain

$$\left\langle \eta_* \frac{d}{dt}, \eta i \right\rangle = \left\langle \frac{d\theta}{dt}\left(q_* \frac{\partial}{\partial\theta}\right) + \frac{d\varphi}{dt}\left(q_* \frac{\partial}{\partial\varphi}\right) + \frac{d\mu}{dt}\left(q_* \frac{\partial}{\partial\mu}\right), q\big(\theta(t), \varphi(t), \mu(t)\big) i \right\rangle$$
$$= \frac{d\mu(t)}{dt} + \frac{1}{2}(1 - \sin\theta(t))\frac{d\varphi(t)}{dt}.$$

This leads us to the desired result.   □

We denoted by Φ a local chart of S^2. Recalling its definition we can extend it to a map $\mathbb{R} \times \mathbb{R} \to S^2$ and can also consider it as a map $S^1 \times S^1 \to S^2$ by identifying S^1 with $\mathbb{R}/2\pi\mathbb{Z}$. From now on we do not distinguish them.

Example 2.1. Let $\gamma : \mathbb{R} \to S^2$ be either a small circle or a great circle on S^2 parameterized by its arclength. It is expressed as

$$\gamma(t) = \sin\theta_0 \cdot i + \cos\theta_0 \cos t \cdot j + \cos\theta_0 \sin t \cdot k$$
$$= \big(j\ -k\ i\big) \begin{pmatrix} \cos\theta_0 \cos t \\ -\cos\theta_0 \sin t \\ \sin\theta_0 \end{pmatrix},$$

with a fixed parameter $\theta_0 \in \mathbb{R}$ ($0 \leq \theta_0 < \pi/2$). That is, the corresponding curve on $(-\pi, \pi) \times (-\pi/2, \pi/2)$ is given by $t \mapsto (\theta(t), \varphi(t)) = (\theta_0, -t)$. Hence the solution $\mu(t)$ of the differential equation (7) is given by

$$\mu(t) = \frac{1}{2}(1 - \sin\theta_0)t + c_0$$

with some constant c_0 in this case. Thus, every horizontal lift of γ is given in the following form:

$$\eta(t) = \left\{\sin\left(\frac{\theta_0}{2}+\frac{\pi}{4}\right) + j\cos\left(\frac{\theta_0}{2}+\frac{\pi}{4}\right)(-i)e^{-it}\right\}\exp\frac{i}{2}\left\{(1-\sin\theta_0)t+c_0\right\}$$

$$= \sin\left(\frac{\theta_0}{2}+\frac{\pi}{4}\right)\exp\frac{i}{2}\left\{(1-\sin\theta_0)t+c_0\right\}$$

$$+ \left\{j\cos\left(\frac{\theta_0}{2}+\frac{\pi}{4}\right)\right\}(-i)\exp\frac{-i}{2}\left\{(1+\sin\theta_0)t+c_0\right\}$$

$$= \sin\left(\frac{\theta_0}{2}+\frac{\pi}{4}\right)\exp\frac{i}{2}\left\{\left(\cos\frac{\theta_0}{2}-\sin\frac{\theta_0}{2}\right)^2 t+c_0\right\}$$

$$+ \frac{1}{\sqrt{2}}\left\{j\cos\left(\frac{\theta_0}{2}+\frac{\pi}{4}\right)\right\}(-i)\exp\frac{-i}{2}\left\{\left(\cos\frac{\theta_0}{2}+\sin\frac{\theta_0}{2}\right)^2 t+c_0\right\}$$

$$= A\,\exp i\left(B^2 t+\frac{c_0}{2}\right) + jB(-i)\,\exp -i\left(A^2 t+\frac{c_0}{2}\right),$$

where $A = \sin\left(\frac{\theta_0}{2}+\frac{\pi}{4}\right)$ and $B = \cos\left(\frac{\theta_0}{2}+\frac{\pi}{4}\right)$.

Example 2.2. The image of a Viviani curve in S^2 is an intersection of a cylinder and S^2 in $\mathbb{R}^3$. We take a smooth curve $\gamma : \mathbb{R} \to \mathbb{R}^3$ defined by

$$\gamma(t) = \sin\frac{t}{2}\,i + j\cos\frac{t}{2}\,e^{it/2} = (j\ -k\ i)\begin{pmatrix}\cos^2\frac{t}{2}\\ -\cos\frac{t}{2}\sin\frac{t}{2}\\ \sin\frac{t}{2}\end{pmatrix}.$$

If we put

$$x(t) = \cos^2\frac{t}{2},\quad y(t) = -\cos\frac{t}{2}\sin\frac{t}{2},\quad z(t) = \sin\frac{t}{2},$$

then we find

$$x^2(t) + y^2(t) + z^2(t) = 1 \quad\text{and}\quad \left(x(t)-\frac{1}{2}\right)^2 + y(t)^2 = \frac{1}{4}.$$

Thus, we find that γ is a Viviani curve in S^2 and that its minimal period is 4π. The corresponding curve of γ in $(-\pi/2,\pi/2)\times(-\pi,\pi)$ is $t \mapsto (\theta(t),\varphi(t)) = (t/2,t/2)$. Hence it is extended to $\mathbb{R} \ni t \mapsto (t/2,t/2) \in \mathbb{R}\times\mathbb{R}$. In this case, the differential equation (7) turns to

$$\frac{d\mu(t)}{dt} + \frac{1}{4}\left(1-\sin\frac{t}{2}\right) = 0,$$

hence we obtain

$$\mu(t) = -\frac{t}{4} - \frac{1}{2}\cos\frac{t}{2} + c_0,$$

with some constant $c_0 \in \mathbb{R}$. Thus, every horizontal lift of this Viviani curve is given in the following form:

$$
\begin{aligned}
\eta(t) &= q\left(\frac{t}{2}, \frac{t}{2}, -\frac{t}{4} - \frac{1}{2}\cos\frac{t}{2} + c_0\right) \\
&= \left\{\left(\sin\frac{t+\pi}{4}\right) + j\left(\cos\frac{t+\pi}{4}\right)(-i)e^{it/2}\right\} \exp i\left(-\frac{t}{4} - \frac{1}{2}\cos\frac{t}{2} + c_0\right) \\
&= \left\{\left(\sin\frac{t+\pi}{4}\right) - j\left(\cos\frac{t+\pi}{4}\right)e^{i(t+\pi)/2}\right\} \exp i\left(-\frac{t}{4} - \frac{1}{2}\cos\frac{t}{2} + c_0\right).
\end{aligned}
$$

The minimal period of this lift is 8π. Hence the image of a horizontal lift of our Viviani curve is a double covering of the image of our Viviani curve. Setting a new (independent) variable ϑ by $\vartheta = (t+\pi)/4$, we find that this horizontal lift is rewritten as

$$
\begin{aligned}
\Xi(\vartheta) &= \eta(t(\vartheta)) = \eta(4\vartheta - \pi) \\
&= \left(\sin\vartheta - j\cos\vartheta\ e^{2i\vartheta}\right) \exp i\left(-\vartheta - \frac{1}{2}\sin(2\vartheta) + \frac{\pi}{4} + c_0\right).
\end{aligned}
$$

3. Hopf tori

In this section, we give parametrizations of Hopf tori of for curves in Examples 2.1 and 2.2, and compute their mean curvatures. It is known that the mean curvature of the Hopf torus of a given curve in S^2 coincides with the curvature of this curve. (see [3]). It is also well-known that every Hopf surface is a flat surface with respect to the metric induced from that on S^3 ([3]).

Example 3.1. Let γ be either a small circle or a great circle in S^2 which is expressed as $\gamma(t) = \Phi(\theta_0,\ t)$ with some θ_0 $(0 \le \theta_0 < \frac{\pi}{2})$ (see Example 2.1). Considering its inverse image by use of (1), the Hopf torus obtained by γ is given by $\Psi_0 : \mathbb{R} \times \mathbb{R} \to S^3$ which is expressed as

$$
\Psi_0(t, u) = \left\{\sin\left(\frac{\theta_0}{2} + \frac{\pi}{4}\right) + j\cos\left(\frac{\theta_0}{2} + \frac{\pi}{4}\right)(-i)e^{-it}\right\}e^{iu}.
$$

Since $\gamma(t+2\pi) = \gamma(t)$, it induces a map $\Psi_0 : S^1 \times S^1 \cong (\mathbb{R}/2\pi\mathbb{Z}) \times (\mathbb{R}/2\pi\mathbb{Z}) \to S^3$. On the other hand, by using the expressions of horizontal lifts of γ, this Hopf torus is represented by $\Psi_1 : \mathbb{R} \times \mathbb{R} \to S^3$ which is expressed as

$$
\Psi_1(t, \mu) = \eta(t)e^{i\mu} = \left\{A\ e^{iB^2 t} + jB(-i)\ e^{-iA^2 t}\right\}e^{i\mu}
$$

with

$$
A = \sin\left(\frac{\theta_0}{2} + \frac{\pi}{4}\right), \quad B = \cos\left(\frac{\theta_0}{2} + \frac{\pi}{4}\right).
$$

This also induces $\Psi_1 : S^1 \times S^1 \cong (\mathbb{R}/2\pi\mathbb{Z}) \times (\mathbb{R}/2\pi\mathbb{Z}) \to S^3$. When the ratio A^2/B^2 is rational, each horizontal lift of this circle is periodic. On contrary, when the ratio A^2/B^2 is irrational, each horizontal lift of this circle forms a dense subset of the Hopf torus by Kronecker's approximation theorem.

Proposition 3.1. *The Hopf surface in Example 3.1 is flat, and its mean curvature H is given by $H = \tan\theta_0$.*

Proof. We use the immersion $\Psi_1 : S^1 \times S^1 \to S^3$ given in Example 3.1. Since $0 \leq \theta_0 < \pi/2$ we see $AB = (\cos\theta_0)/2 \neq 0$. We set

$$e_1 = \frac{1}{AB}\frac{\partial}{\partial t}, \quad e_2 = \frac{\partial}{\partial\mu}.$$

The $SO(4)$–frame field $g(t_1, \mu)$ along the immersion Ψ_1 is given by

$$g = g(t, \mu) = \big(\Psi_1, \Psi_{1*}(e_1), \Psi_{1*}(e_2), \xi\big)$$

where

$$\Psi_{1*}(e_1) = \big\{iB\ e^{iB^2 t} - jA\ e^{-iA^2 t}\big\}e^{i\mu},$$

$$\Psi_{1*}(e_2) = \big\{iA\ e^{iB^2 t} + jB\ e^{-iA^2 t}\big\}e^{i\mu}\ (= \Psi_1 i),$$

$$\xi = \big\{B\ e^{iB^2 t} + jA\ i\ e^{-iA^2 t}\big\}e^{i\mu}.$$

The Maurer Cartan form Ω is given by

$$\Omega = g^{-1}dg$$

$$= \begin{pmatrix} 0 & -\langle\Psi_{1*}(e_1), d\Psi_1\rangle & -\langle\Psi_{1*}(e_2), d\Psi_1\rangle & 0 \\ \langle\Psi_{1*}(e_1), d\Psi_1\rangle & 0 & \langle\Psi_{1*}(e_1), d(\Psi_{1*}(e_2))\rangle & \langle\Psi_{1*}(e_1), d\xi\rangle \\ \langle\Psi_{1*}(e_2), d\Psi_1\rangle & \langle\Psi_{1*}(e_2), d(\Psi_{1*}(e_1))\rangle & 0 & \langle\Psi_{1*}(e_2), d\xi\rangle \\ 0 & \langle\xi, d(\Psi_{1*}(e_1))\rangle & \langle\xi, d(\Psi_{1*}(e_2))\rangle & 0 \end{pmatrix}$$

$$= \begin{pmatrix} 0 & -\omega_1 & -\omega_2 & 0 \\ \omega_1 & 0 & \omega_{12} & \omega_{13} \\ \omega_2 & \omega_{21} & 0 & \omega_{23} \\ 0 & \omega_{31} & \omega_{32} & 0 \end{pmatrix},$$

where $\omega_1 = ABdt_1$, $\omega_2 = d\mu$ and

$$\omega_{12} = -\omega_{21} = 0,$$

$$\omega_{13} = -\omega_{31} = -\sin\theta_0 dt_1 + d\mu,$$

$$\omega_{23} = -\omega_{32} = \omega_1 = ABdt_1.$$

Therefore we obtain

$$\begin{pmatrix} \omega_{13} \\ \omega_{23} \end{pmatrix} = \begin{pmatrix} -\sin\theta_0/AB & 1 \\ 1 & 0 \end{pmatrix} \begin{pmatrix} \omega_1 \\ \omega_2 \end{pmatrix}.$$

From this the shape operator A_ξ with respect to the orthonormal base $\{e_1, e_2\}$ is given by

$$A_\xi = \begin{pmatrix} \sin\theta_0/AB & -1 \\ -1 & 0 \end{pmatrix}.$$

The mean curvature of this surface is given by

$$H = \frac{1}{2}\mathrm{tr}A_\xi = \frac{\sin\theta_0}{2AB} = \tan\theta_0.$$

Also since the connection 1-form ω_{12} vanishes, the Gauss curvature is identically 0.   □

Example 3.2. Let γ be a Viviani curve given by $\gamma(t) = \Phi(t/2, t/2)$ in Example 2.2. Considering its inverse image by use of (1), the Hopf torus obtained by γ is given by $\mathcal{H}_0 : \mathbb{R} \times \mathbb{R} \to S^3$ which is wich is expressed as

$$\mathcal{H}_0(t,\mu) = \left(\sin\left(\frac{t+\pi}{4}\right) + j\cos\left(\frac{t+\pi}{4}\right)e^{i(t-\pi)/2}\right) e^{i\mu}.$$

As $\gamma(t + 4\pi) = \gamma(t)$, it induces $\mathcal{H}_0 : S^1 \times S^1 \cong \mathbb{R}/4\pi\mathbb{Z} \times \mathbb{R}/2\pi\mathbb{Z} \to S^3$. If we put $\vartheta = (t+\pi)/4$, as $(t-\pi)/2 = 2\vartheta - \pi$, we have

$$\mathcal{H}_1(\vartheta, \mu) = \left(\sin\vartheta - j\cos\vartheta e^{2i\vartheta}\right) e^{i\mu}, \tag{8}$$

Then this map $\mathcal{H}_1 : \mathbb{R} \times \mathbb{R} \to S^3$ has the following periodicities:

$$\mathcal{H}_1(\vartheta + \pi, \mu + \pi) = \mathcal{H}_1(\vartheta, \mu), = \mathcal{H}_1(\vartheta, \mu + 2\pi).$$

In particular, it satisfies $\mathcal{H}_1\left(\pi/4, \mu\right) = \mathcal{H}_1\left(5\pi/4, \mu + \pi\right)$ for each $\mu \in \mathbb{R}$. Thus, the fundamental domain of the immersion $\mathcal{H}_1$ is the parallelogram

$$\Pi_1 = \left\{(\vartheta, \mu) \in \mathbb{R}^2 \;\middle|\; \frac{\pi}{4} \leq \vartheta < \frac{5\pi}{4}, \;\; \vartheta + \frac{\pi}{4} \leq \mu < \vartheta + \frac{5\pi}{4} \right\}. \tag{9}$$

We note that the set

$$\ell_1 = \left\{ \left(\frac{3\pi}{4}, \mu\right) \in \mathbb{R}^2 \;\middle|\; \frac{3\pi}{4} \leq \mu < \frac{11\pi}{4} \right\}$$

is the inverse image of the self-intersection point of our Viviani curve in S^2, which corresponds to the center vertical line in Fig. 1. The induced metric $g_1 = \mathcal{H}_1^*\langle\,,\,\rangle_{S^3}$ is given by

$$g_1 = (1 + 4\cos^2\vartheta)d\vartheta^2 + 2\cos^2\vartheta d\vartheta \odot d\mu + d\mu^2.$$

Let ρ_1 and ρ_2 be curves showing parts of the boundary of the fundamental domain Π_1 defined by

$$\rho_1(\vartheta) = (\vartheta, \vartheta) \quad \text{and} \quad \rho_2(\vartheta) = (\vartheta, \vartheta + 2\pi) \qquad (\pi/4 \leq \vartheta \leq 5\pi/4).$$

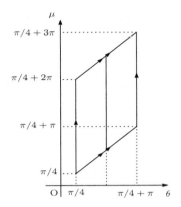

Fig. 1.   The fundamental domain of the immersion $\mathcal{H}_1$

Then these curves ρ_1, ρ_2 are not geodesics in $\mathcal{H}_1(\mathbb{R}^2)$ with respect to g_1. We represent our Hopf torus by using the expressions of horizontal lifts of γ. As we see in Example 2.2, every horizontal lift of γ is expressed as $\varXi$ with some c_0. Since we use c_0 to show a point passing through, by taking $c_0 = -\pi/4$, we see that our Hopf torus is represented by $\mathcal{H}_2 : \mathbb{R} \times \mathbb{R} \to S^3$ defined by

$$\mathcal{H}_2(\vartheta, \varphi) = \varXi(\vartheta)\, e^{i\varphi} = \left(\sin \vartheta - j \cos \vartheta \; e^{2i\vartheta}\right) \exp i\left(\varphi - \vartheta - \frac{1}{2}\sin 2\vartheta\right).$$

One can easily check that the induced metric g_2 on $\mathbb{R}^2$ with respect to the immersion $\mathcal{H}_2$ is given by

$$g_2 = (1 + \sin^2 2\vartheta)d\vartheta^2 + d\varphi^2, \tag{10}$$

Then, for a fixed $\varphi_0 \in \mathbb{R}$, the curve $\vartheta \mapsto \mathcal{H}_2(\vartheta, \varphi_0)$ is a geodesic with respect to this metric g_2.

Proposition 3.2. *The Hopf hypersurface in Example 3.2 is flat, and its mean curvature is given by*

$$H = \frac{\cos 2\vartheta \; (2 + \sin^2 2\vartheta)}{1 + \sin^2 2\vartheta}.$$

Proof. We use the immersion $\mathcal{H}_1 : \mathbb{R}^2 \to S^2$ given in Example 3.2. We set

$$e_1 = \frac{1}{\sqrt{1 + \sin^2 2\vartheta}} \frac{\partial}{\partial \vartheta}, \quad e_2 = \frac{\partial}{\partial \varphi}.$$

Then (e_1, e_2) is a local orthonormal frame field on $\mathbb{R}^2$ with respect to this metric g_2. The $SO(4)$-moving local frame field $g = g(\vartheta, \varphi)$ along the immersion $\mathcal{H}_2$ is given by

$$g = \big(\mathcal{H}_2, \mathcal{H}_{2*}(e_1), \mathcal{H}_{2*}(e_2), \xi\big),$$

where

$$\mathcal{H}_{2*}(e_1) = \mathcal{H}_{2*}\left(\frac{1}{\sqrt{1+\sin^2 2\vartheta}}\frac{\partial}{\partial\vartheta}\right) = \frac{1}{\sqrt{1+\sin^2 2\vartheta}}\frac{d(\Xi(\vartheta))}{d\vartheta}e^{i\varphi}$$

$$= \frac{1}{\sqrt{1+\sin^2 2\vartheta}}$$

$$\times \Big\{\big(\cos(2\varphi-\sin 2\vartheta) + \sin 2\vartheta\sin(2\varphi-\sin 2\vartheta)\big)\mathcal{H}_2(\vartheta,\varphi)j$$

$$+ \big(-\sin(2\varphi-\sin 2\vartheta) + \sin 2\vartheta\cos(2\varphi-\sin 2\vartheta)\big)\mathcal{H}_2(\vartheta,\varphi)k\Big\},$$

$$\mathcal{H}_{2*}(e_2) = \mathcal{H}_{2*}\big(\frac{\partial}{\partial\varphi}\big) = \mathcal{H}_2(\vartheta,\varphi)i,$$

$$\xi(\vartheta,\varphi) = \mathcal{H}_{2*}(e_1)i$$

$$= \frac{1}{\sqrt{1+\sin^2 2\vartheta}}$$

$$\times \Big\{\big(-\sin(2\varphi-\sin 2\vartheta) + \sin 2\vartheta\cos(2\varphi - \sin 2\vartheta)\big)\mathcal{H}_2(\vartheta,\varphi)j$$

$$- \big(\cos(2\varphi - \sin 2\vartheta) + \sin 2\vartheta\sin(2\varphi - \sin 2\vartheta)\big)\mathcal{H}_2(\vartheta,\varphi)k\Big\}.$$

In order to check that this g is an $SO(4)$-moving local frame field, we need the following.

Claim 1.

$$\left\langle \mathcal{H}_{2*}\left(\frac{\partial}{\partial\vartheta}\right), \mathcal{H}_2(\vartheta,\varphi)j \right\rangle = \cos(2\varphi-\sin 2\vartheta) + \sin 2\vartheta\sin(2\varphi-\sin 2\vartheta),$$

$$\left\langle \mathcal{H}_{2*}\left(\frac{\partial}{\partial\varphi}\right), \mathcal{H}_2(\vartheta,\varphi)k \right\rangle = -\sin(2\varphi-\sin 2\vartheta) + \sin 2\vartheta\cos(2\varphi-\sin 2\vartheta).$$

We here compute right-hand sides of these equalities in Claim 1. Since, $\Xi(\vartheta)$ is a horizontal lift of γ, we have

$$\frac{d\Xi}{d\vartheta} = \left\langle\frac{d\Xi}{d\vartheta},\Xi j\right\rangle\Xi j + \left\langle\frac{d\Xi}{d\vartheta},\Xi k\right\rangle\Xi k.$$

From this we get

$$\left\langle \mathcal{H}_{2*}\left(\frac{\partial}{\partial\vartheta}\right), \mathcal{H}_2(\vartheta,\varphi)j \right\rangle = \left\langle\frac{d\Xi}{d\vartheta}e^{i\varphi}, \Xi\, e^{i\varphi}j\right\rangle$$

$$= \left\langle\frac{d\Xi}{d\vartheta}, \Xi j\right\rangle\left\langle\Xi j\, e^{i\varphi},\Xi\, e^{i\varphi}j\right\rangle + \left\langle\frac{d\Xi}{d\vartheta}, \Xi k\right\rangle\left\langle\Xi k\, e^{i\varphi}, \Xi\, e^{i\varphi}j\right\rangle$$

$$= \Big\langle \frac{d\Xi}{d\vartheta}, \ \Xi j \Big\rangle \big\langle j \ e^{i\varphi}, \ e^{i\varphi} j \big\rangle + \Big\langle \frac{d\Xi}{d\vartheta}, \ \Xi k \Big\rangle \big\langle k \ e^{i\varphi}, \ e^{i\varphi} j \big\rangle$$

$$= \Big\langle \frac{d\Xi}{d\vartheta}, \ \Xi j \Big\rangle \big\langle e^{-i\varphi} j, \ e^{i\varphi} j \big\rangle + \Big\langle \frac{d\Xi}{d\vartheta}, \ \Xi k \Big\rangle \big\langle e^{-i\varphi} k, \ e^{i\varphi} j \big\rangle$$

$$= \cos 2\varphi \ \Big\langle \frac{d\Xi}{d\vartheta}, \ \Xi j \Big\rangle + \sin 2\varphi \ \Big\langle \frac{d\Xi}{d\vartheta}, \ \Xi k \Big\rangle.$$

Similarly, we get

$$\Big\langle \mathcal{H}_{2*}\Big(\frac{\partial}{\partial \vartheta}\Big), \mathcal{H}_2(\vartheta, \varphi) k \Big\rangle = - \sin 2\varphi \ \Big\langle \frac{d\Xi}{d\vartheta}, \ \Xi j \Big\rangle + \cos 2\varphi \ \Big\langle \frac{d\Xi}{d\vartheta}, \ \Xi k \Big\rangle.$$

Since

$$\frac{d\Xi}{d\vartheta} = \Big(\cos \vartheta + j(\sin \vartheta - 2i \cos \vartheta) \ e^{2i\vartheta}\Big) \exp i\Big(-\vartheta - \frac{1}{2}\sin 2\vartheta\Big)$$

$$- \Big(\sin \vartheta - j \cos \vartheta \ e^{2i\vartheta}\Big) i(1 + \sin 2\vartheta) \exp i\Big(-\vartheta - \frac{1}{2}\sin 2\vartheta\Big)$$

$$= \Big(\cos \vartheta + j(\sin \vartheta - 2i \cos \vartheta) \ e^{2i\vartheta}\Big) \exp i\Big(-\vartheta - \frac{1}{2}\sin 2\vartheta\Big)$$

$$- (1 + \sin 2\vartheta)\Xi(\vartheta) i,$$

we find

$$\Big\langle \frac{d\Xi}{d\vartheta}, \ \Xi j \Big\rangle = \Big\langle \cos \vartheta + j(\sin \vartheta - 2i \cos \vartheta) \ e^{2i\vartheta}, (\sin \vartheta - j \cos \vartheta) j e^{i \sin 2\vartheta} \Big\rangle$$

$$= \cos(\sin 2\vartheta) - \sin 2\vartheta \cdot \sin(\sin 2\vartheta),$$

$$\Big\langle \frac{d\Xi}{d\vartheta}, \ \Xi k \Big\rangle = \sin(\sin 2\vartheta) + \sin 2\vartheta \cdot \cos(\sin 2\vartheta).$$

We hence obtain

$$\Big\langle \mathcal{H}_{2*}\Big(\frac{\partial}{\partial \vartheta}\Big), \mathcal{H}_2(\vartheta, \varphi) j \Big\rangle = \cos 2\varphi \ \big(\cos(\sin 2\vartheta) - \sin 2\vartheta \cdot \sin(\sin 2\vartheta)\big)$$

$$+ \sin 2\varphi \ \big(\sin(\sin 2\vartheta) + \sin 2\vartheta \cdot \cos(\sin 2\vartheta)\big)$$

$$= \cos(2\varphi - \sin 2\vartheta) + \sin 2\vartheta \sin(2\varphi - \sin 2\vartheta),$$

$$\Big\langle \mathcal{H}_{2*}\Big(\frac{\partial}{\partial \vartheta}\Big), \mathcal{H}_2(\vartheta, \varphi) k \Big\rangle = - \sin(2\varphi - \sin 2\vartheta) + \sin 2\vartheta \cos(2\varphi - \sin 2\vartheta),$$

and get the assertion of Claim 1.

By using the local frame field g we compute the Maurer Cartan form

$$\Omega = {}^t g dg = \begin{pmatrix} 0 & -\omega_1 & -\omega_2 & 0 \\ \omega_1 & 0 & \omega_{12} & \omega_{13} \\ \omega_2 & \omega_{21} & 0 & \omega_{23} \\ 0 & \omega_{31} & \omega_{32} & 0 \end{pmatrix}.$$

Claim 2. The entries of the Maurer Cartan form Ω are as follows:

$$\omega_1 = \sqrt{1 + \sin^2 2\vartheta}\, d\vartheta, \qquad \omega_2 = d\varphi, \qquad \omega_{12} = -\omega_{21} = 0,$$

$$\omega_{13} = -\omega_{31} = -\frac{2\cos 2\vartheta(2 + \sin^2 2\vartheta)}{1 + \sin^2 2\vartheta}\, d\vartheta - d\varphi$$

$$= -\frac{2\cos 2\vartheta(2 + \sin^2 2\vartheta)}{(1 + \sin^2 2\vartheta)^{3/2}}\omega_1 - \omega_2,$$

$$\omega_{23} = -\omega_{32} = -\sqrt{1 + \sin^2 2\vartheta}\, d\vartheta = -\omega_1.$$

We here computer these entries. Since $g_2 = (1 + \sin^2 2\vartheta)d\vartheta^2 + d\varphi^2$, we have $\omega_1 = \sqrt{1 + \sin^2 2\vartheta}\, d\vartheta$ and $\omega_2 = d\varphi$. As

$$\omega_{12} = \langle \mathcal{H}_{2*}(e_1), d\mathcal{H}_{2*}(e_2) \rangle$$

$$= \left\langle \mathcal{H}_{2*}(e_1), \frac{\partial}{\partial\vartheta}\left(\mathcal{H}_{2*}(e_2)\right) \right\rangle d\vartheta + \left\langle \mathcal{H}_{2*}(e_1), \frac{\partial}{\partial\varphi}\left(\mathcal{H}_{2*}(e_2)\right) \right\rangle d\varphi,$$

and

$$\left\langle \mathcal{H}_{2*}(e_1), \frac{\partial}{\partial\vartheta}\left(\mathcal{H}_{2*}(e_2)\right) \right\rangle = \left\langle \mathcal{H}_{2*}(e_1), \mathcal{H}_{2*}\left(\frac{\partial}{\partial\vartheta}\right)i \right\rangle$$

$$= \left\langle \mathcal{H}_{2*}(e_1), \sqrt{1 + \sin^2 2\vartheta}\left(\mathcal{H}_{2*}(e_1)\right)i \right\rangle$$

$$= \left\langle \mathcal{H}_{2*}(e_1), \sqrt{1 + \sin^2 2\vartheta}\,\xi(\vartheta, \varphi) \right\rangle = 0,$$

$$\left\langle \mathcal{H}_{2*}(e_1), \frac{\partial}{\partial\varphi}\left(\mathcal{H}_{2*}(e_2)\right) \right\rangle = \left\langle \mathcal{H}_{2*}(e_1), \mathcal{H}_{2*}\left(\frac{\partial}{\partial\varphi}\right)i \right\rangle$$

$$= \left\langle \mathcal{H}_{2*}(e_1), (\mathcal{H}_2(\vartheta, \varphi)i)i \right\rangle = \left\langle \mathcal{H}_{2*}(e_1), -\mathcal{H}_2(\vartheta, \varphi) \right\rangle = 0,$$

we get $\omega_{12} = -\omega_{21} = 0$.

Next we compute ω_{13}. We have

$$\omega_{13} = -\langle \xi, d\mathcal{H}_{2*}(e_1) \rangle$$

$$= -\left\langle \mathcal{H}_{2*}(e_1)i, \frac{\partial}{\partial\vartheta}\left(\mathcal{H}_{2*}(e_1)\right) \right\rangle d\vartheta - \left\langle \mathcal{H}_{2*}(e_1)i, \frac{\partial}{\partial\varphi}\left(\mathcal{H}_{2*}(e_1)\right) \right\rangle d\varphi.$$

By using the canonical complex structure J_{S^2} on S^2, we can compute the inner product in the first term as

$$\left\langle \mathcal{H}_{2*}(e_1)i, \frac{\partial}{\partial\vartheta}\left(\mathcal{H}_{2*}(e_1)\right) \right\rangle$$

$$= \left\langle \frac{1}{\sqrt{1 + \sin^2 2\vartheta}}\frac{d\Xi}{d\vartheta}e^{i\varphi}i, \left(\frac{1}{\sqrt{1 + \sin^2 2\vartheta}}\frac{d^2\Xi}{d\vartheta^2}\right)e^{i\varphi} \right\rangle$$

$$= \frac{1}{1 + \sin^2 2\vartheta} \left\langle \frac{d\Xi}{d\vartheta} i, \ \frac{d^2\Xi}{d\vartheta^2} \right\rangle$$

$$= \frac{1}{1 + \sin^2 2\vartheta} \left\langle \nabla^{S^3}_{\frac{d}{d\vartheta}} \Xi_* \left(\frac{d}{d\vartheta} \right), \ \Xi_* \left(\frac{d}{d\vartheta} \right) i \right\rangle$$

$$= \frac{1}{4(1 + \sin^2 2\vartheta)} \left\langle (\pi \circ \Xi)_* \left(\nabla^{S^2}_{\frac{d}{d\vartheta}} \frac{d}{d\vartheta} \right), \ J_{S^2}(\pi \circ \Xi)_* \left(\frac{d}{d\vartheta} \right) \right\rangle_{S^2}$$

$$= \left\langle \frac{d^2}{d\vartheta^2} (\pi \circ \Xi), \ \frac{d}{d\vartheta} (\pi \circ \Xi) \times (\pi \circ \Xi) \right\rangle.$$

Here, as we have

$$(\pi \circ \Xi)(\vartheta) = \Xi(\vartheta) i \overline{\Xi(\vartheta)} = -i \cos 2\vartheta + j \sin 2\vartheta \sin 2\vartheta + k \sin 2\vartheta \cos 2\vartheta$$

$$= \left(j \ -k \ i \right) \begin{pmatrix} \sin 2\vartheta \sin 2\vartheta \\ -\sin 2\vartheta \cos 2\vartheta \\ -\cos 2\vartheta \end{pmatrix},$$

we obtain

$$\left\langle \mathcal{H}_{2*}(e_1) i, \ \frac{\partial}{\partial \vartheta} (\mathcal{H}_{2*}(e_1)) \right\rangle$$

$$= \frac{1}{4(1 + \sin^2 2\vartheta)} 8 \cos 2\vartheta (2 + \sin^2 2\vartheta) = \frac{2 \cos 2\vartheta (2 + \sin^2 2\vartheta)}{1 + \sin^2 2\vartheta}.$$

similarly, we can compute the inner product in second term as

$$\left\langle \mathcal{H}_{2*}(e_1) i, \ \frac{\partial}{\partial \varphi} (\mathcal{H}_{2*}(e_1)) \right\rangle = \left\langle \mathcal{H}_{2*}(e_1) i, \ \frac{1}{\sqrt{1 + \sin^2 2\vartheta}} \nabla^{S^3}_{\frac{\partial}{\partial \varphi}} \mathcal{H}_{2*} \left(\frac{\partial}{\partial \vartheta} \right) \right\rangle$$

$$= \left\langle \mathcal{H}_{2*}(e_1) i, \ \frac{1}{\sqrt{1 + \sin^2 2\vartheta}} \nabla^{S^3}_{\frac{\partial}{\partial \vartheta}} \mathcal{H}_{2*} \left(\frac{\partial}{\partial \varphi} \right) \right\rangle$$

$$= \left\langle \mathcal{H}_{2*}(e_1) i, \ \frac{1}{\sqrt{1 + \sin^2 2\vartheta}} \nabla^{S^3}_{\frac{\partial}{\partial \vartheta}} (\mathcal{H}_2(\vartheta, \varphi) i) \right\rangle$$

$$= \left\langle \mathcal{H}_{2*}(e_1) i, \ \frac{1}{\sqrt{1 + \sin^2 2\vartheta}} \mathcal{H}_{2*} \left(\frac{\partial}{\partial \vartheta} \right) i \right\rangle$$

$$= \left\langle \mathcal{H}_{2*}(e_1) i, \ \mathcal{H}_{2*}(e_1) i \right\rangle = 1.$$

Hence we find

$$\omega_{13} = -\omega_{31} = -\frac{2 \cos 2\vartheta (2 + \sin^2 2\vartheta)}{1 + \sin^2 2\vartheta} d\vartheta - d\varphi.$$

Lastly we have

$$
\begin{aligned}
\omega_{23} &= -\big\langle \xi,\ d\mathcal{H}_{2*}(e_2) \big\rangle \\
&= -\Big\langle \mathcal{H}_{2*}(e_1)i,\ \frac{\partial}{\partial\vartheta}\big(\mathcal{H}_{2*}(e_2)\big) \Big\rangle d\vartheta - \Big\langle \mathcal{H}_{2*}(e_1)i,\ \frac{\partial}{\partial\varphi}\big(\mathcal{H}_{2*}(e_2)\big) \Big\rangle d\varphi \\
&= -\Big\langle \mathcal{H}_{2*}(e_1)i,\ \frac{\partial}{\partial\vartheta}\big(\mathcal{H}_2(\vartheta,\varphi)i\big) \Big\rangle d\vartheta - \Big\langle \mathcal{H}_{2*}(e_1)i,\ \frac{\partial}{\partial\varphi}\big(\mathcal{H}_2(\vartheta,\varphi)i\big) \Big\rangle d\varphi \\
&= -\big\langle \mathcal{H}_{2*}(e_1)i,\ \sqrt{1+\sin^2 2\vartheta}\,\mathcal{H}_{2*}(e_1)i \big\rangle d\vartheta \\
&\quad - \big\langle \mathcal{H}_{2*}(e_1)i,\ (\mathcal{H}_2(\vartheta,\varphi)i)\,i \big\rangle d\varphi \\
&= -\sqrt{1+\sin^2 2\vartheta}\,d\vartheta = -\omega_1.
\end{aligned}
$$

This completes our computation for Claim 2.

By Claim 2, we find that the shape operator A_ξ with respect to the orthonormal base $\{e_1,\ e_2\}$ is given by

$$
A_\xi = \begin{pmatrix} \dfrac{2\cos 2\vartheta(2+\sin^2 2\vartheta)}{(1+\sin^2 2\vartheta)} & 1 \\ 1 & 0 \end{pmatrix}.
$$

Hence the mean curvature of our surface is

$$
H = \frac{1}{2}\mathrm{tr} A_\xi = \frac{\cos 2\vartheta(2+\sin^2 2\vartheta)}{(1+\sin^2 2\vartheta)}
$$

and the Gauss curvature vanishes.   □

4. On metric on the Hopf torus of a Viviani curve

In this section we study more on our Hopf torus for a Viviani curve given in Example 3.2. As we gave metrics g_1, g_2 corresponding to the immersions $\mathcal{H}_1$ and $\mathcal{H}_2$, we shall give fundamental domains of this torus whose boundaries are formed by geodesic segments with respect to g_1 and g_2. As we see in Example 3.2 the map $\mathcal{H}_1 : \mathbb{R}^2 \to S^3$ given by

$$
\mathcal{H}_1(\vartheta,\mu) = \Big(\sin\vartheta - j\cos\vartheta e^{2i\vartheta}\Big)e^{i\mu}
$$

induces a map $\mathcal{H}_1 : \mathbb{R}/\pi\mathbb{Z} \times \mathbb{R}/2\pi\mathbb{Z} \to S^3$, which is a representation of the Hopf torus related to our Viviani curve. The induced metric on the Hopf torus $\mathcal{H}_1(T^2)$ was given by

$$
g_1 = \mathcal{H}_1^*\langle\ ,\ \rangle_{S^3} = (1+4\cos^2\vartheta)d\vartheta^2 + 4\cos^2\vartheta d\vartheta \odot d\mu + d\mu^2.
$$

Recalling Example 3.2 we know that two curves ρ_1, ρ_2 showing parts of the boundary of the fundamental domain Π_1 in $\mathbb{R}^2$ for $\mathcal{H}_1$ are not geodesics with respect to g_1. The center dotted curve in Fig. 2 which is not a straight line is a geodesic with respect to g_1.

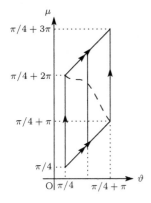

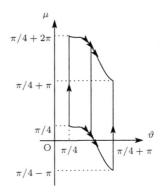

Fig. 2.   The fundamental domain of Π_1 Fig. 3.   The fundamental domain of Π_2

If we change the fundamental domain from Π_1 to Π_2 given by

$$\Pi_2 = \left\{ (\vartheta, \mu) \in R^2 \;\middle|\; \frac{\pi}{4} \leq \vartheta < \frac{5\pi}{4}, \;\; \alpha_0(\vartheta) \leq \mu < \alpha_1(\vartheta) \right\}$$

with

$$\alpha_0(\vartheta) = \frac{\pi + 1}{2} - \vartheta - \frac{1}{2}\sin 2\vartheta \quad \text{and} \quad \alpha_1(\vartheta) = \frac{5\pi + 1}{2} - \vartheta - \frac{1}{2}\sin 2\vartheta$$

as in Fig. 3, then each curve of the boundary is a geodesic.

Our Hopf torus is also represented by $\mathcal{H}_2 : \mathbb{R}^2 \to S^3$ given by

$$\mathcal{H}_2(\vartheta, \varphi) = \left(\sin \vartheta - j\cos \vartheta \; e^{2i\vartheta}\right) \; \exp i\left(\varphi - \vartheta - \frac{1}{2}\sin 2\vartheta\right).$$

This induces a map $\mathcal{H}_2 : S^1 \times S^1 \cong \mathbb{R}/\pi\mathbb{Z} \times \mathbb{R}/2\pi\mathbb{Z} \to S^3$. The induced metric $g_2 = \mathcal{H}_2^*\langle \; , \; \rangle_{S^3}$ on our Hopf torus was given by

$$g_2 = (1 + \sin^2 2\vartheta)d\vartheta^2 + d\varphi^2,$$

hence for each $\varphi_0 \in \mathbb{R}$ the curve $\vartheta \mapsto \mathcal{H}_2(\vartheta, \varphi_0)$ is a geodesic with respect to g_2. We here define a map $F : \mathbb{R}/\pi\mathbb{Z} \times \mathbb{R}/2\pi\mathbb{Z} \to \mathbb{R}/\pi\mathbb{Z} \times \mathbb{R}/2\pi\mathbb{Z}$ by

$$F(\vartheta, \mu) = \left(\vartheta, \mu + \vartheta + \frac{1}{2}\sin 2\vartheta\right).$$

Then it is an isometry of (Π_1, g_1) to $(F(\Pi_1), g_2)$. Through this isometry two lines of the boundary of horizontal part of Π_1 correspond to horizontal

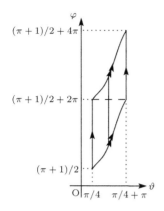

Fig. 4.   The domain $F(\Pi_1)$

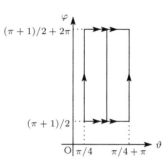

Fig. 5.   The domain $F(\Pi_2)$

curves in Fig. 4. The image $F(\Pi_2)$ is like Fig. 5. The four curves showing the boundary of $F(\Pi_2)$ are geodesics with respect to g_2.

In order to get some topological properties of horizontal lifts of our Viviani curve, we give its image and an image of its projection onto a plane. We take the following horizontal lift of our Viviani curve

$$\mathcal{H}_1(\vartheta, 0) = \sin\vartheta - j\cos\vartheta e^{2i\vartheta}$$

$$= \begin{pmatrix} -j & k & 1 \end{pmatrix} \begin{pmatrix} \cos\vartheta\cos 2\vartheta \\ \cos\vartheta\sin 2\vartheta \\ \sin\vartheta \end{pmatrix} = \begin{pmatrix} -j & k & 1 \end{pmatrix} \begin{pmatrix} x(\vartheta) \\ y(\vartheta) \\ z(\vartheta) \end{pmatrix}$$

(see Fig. 6). Here, we identify $\mathbb{R}^3$ with $\mathrm{span}_{\mathbb{R}}\{-j, k, 1\}$. This curve is included in $S^3 \cap \mathrm{span}_{\mathbb{R}}\{-j, k, 1\}$. Since the image of our Viviani curve has one self-intersection point and the image of its horizontal lift is a double covering, this lift has two self-intersections.

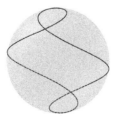

Fig. 6.   A horizontal lift in S^2

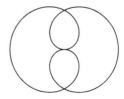

Fig. 7.   The projection of a horizontal lift into $\mathcal{P}$

Since we have

$$4\big\{x(\vartheta)^2 + y(\vartheta)^2\big\}^3 - 4\big\{x(\vartheta)^2 + y(\vartheta)^2\big\}^2 + y(\vartheta)^2 = 0,$$

its projected image onto a plane $\mathcal{P} = \mathrm{span}_{\mathbb{R}}\{-j, k\}$ is like Fig. 7. One can see two self-intersections and a point where the projected curve is tangent to itself. The third point appears by the projection which corresponds to the top and the bottom in Fig. 6. The Hopf torus corresponding to our Viviani curve is obtained by rotating the curve in Fig. 6 under the action of $e^{i\varphi}$ (see Fig. 8).

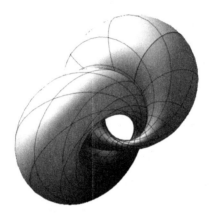

Fig. 8.   The Hopf torus for our Vivuani curve

References

[1] H. Hashimoto and M. Ohashi, *On generalized cylindrical helices and Lagrangian surfaces of $\mathbb{R}^4$*. J. Geom. **106**, 405–420, (2015).

[2] B. O'Neill, *Elementary Differential Geometry*. Revised Second Edition, Elsevier Inc, 2006

[3] U. Pinkall, *Hopf tori in S^3*. Invent. Math. **81**, 379–386, (1985).

Received May 9, 2017
Revised   June 1, 2017

Contemporary Perspectives
in Differential Geometry
and its Related Fields   151 – 159

HOMOTOPY GROUPS OF $G_2/Sp(1)$ AND $G_2/U(2)$

Fuminori NAKATA*

*Faculty of Human Development and Culture, Fukushima University,
Kanayagawa, Fukushima, Japan
E-mail: fnakata@educ.fukushima-u.ac.jp*

The 3rd homotopy groups of homogeneous spaces $G_2/Sp(1)_\pm$ and $G_2/U(2)_\pm$ are determined. Consequently we see that $G_2/Sp(1)_+$ and $G_2/Sp(1)_-$ are not homeomorphic to each other, and similarly for $G_2/U(2)_+$ and $G_2/U(2)_-$. The key is the notion of Dynkin index, which is also explained.

Keywords: Homotopy group; Dynkin index.

1. Introduction

There exist several homogeneous spaces with G_2-symmetry, where G_2 is the 14-dimensional simple Lie group defined as the automorphism group of the Cayley algebra $\mathbb{O}$. In particular, the following diagram is known (see [9])

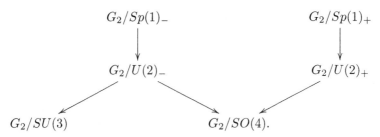

On this diagram, it is known that $G_2/Sp(1)_\pm$ are not homeomorphic to each other. This fact is briefly explained in [3]13.6.3, and we give a detail proof of this fact in this article (Theorem 5.2). By a similar method, we show that $G_2/U(2)_\pm$ are also not homeomorphic to each other (Theorem 5.3). To prove these properties, we determine the third homotopy groups $\pi_3(G_2/Sp(1)_\pm)$ and $\pi_3(G_2/U(2)_\pm)$. The main tool on the proof is the *index* (Dynkin index). In this article, we also explain the definition and some properties of the index.

---
*Partially supported by JSPS Grant-in-Aid for Scientific Research (C) 16K05118.

We remark that the above diagram is related to the twistor theory. It is known that the 8-dimensional homogeneous space $G_2/SO(4)$ is equipped with a structure of a quaternion Kähler Riemannian symmetric space (see [2], [3], [8]). The twistor fibration of this quaternion Kähler structure is $G_2/U(2)_+ \to G_2/SO(4)$. Consequently, it is known that $G_2/Sp(1)_+$ is a 11-dimensional 3-Sasakian manifold. On the other hand, the double fibration $G_2/SU(3) \leftarrow G_2/U(2)_- \to G_2/SO(4)$ is considered as an analogy of the classical Penrose type twistor correspondence (see [9], and [10] for the original Penrose theory). We also remark that $G_2/U(2)_-$ is diffeomorphic to the Grassmannian manifold $G_2(\mathbb{R}^7)$, and $G_2/Sp(1)_-$ is diffeomorphic to the Stiefel manifold $V_2(\mathbb{R}^7)$.

2. G_2 and its subgroups

Let $\mathbb{H} = \{x + yi + zj + wk \mid x, y, z, w \in \mathbb{R}\}$ be the set of quaternions ($i^2 = j^2 = k^2 = -1, ij = -ji = k$), and $Sp(1) \subset \mathbb{H}$ be the group of unit elements. Let $\mathbb{O} = \mathbb{H} \oplus \mathbb{H}\epsilon$ be the Cayley algebra, where the multiplication on $\mathbb{O}$ is $(a + b\epsilon)(c + d\epsilon) = (ac - \bar{d}b) + (da + b\bar{c})\epsilon$. The Lie group G_2 is defined by

$$G_2 = \text{Aut}\mathbb{O} = \{g \in GL(\mathbb{O}) \mid g(xy) = g(x)g(y) \text{ for any } x, y \in \mathbb{O}\}. \quad (1)$$

It is known that the Lie group G_2 is 14-dimensional and simple (see [4]).

There are two embeddings $\rho_\pm : Sp(1) \to G_2$ defined by

$$\rho_+(q)(a + b\epsilon) = a + (qb)\epsilon, \qquad \rho_-(q)(a + b\epsilon) = qa\bar{q} + (b\bar{q})\epsilon \quad (2)$$

(see [5]). We write $Sp(1)_\pm = \rho_\pm(Sp(1))$. We also define subgroups $U(2)_\pm \subset G_2$ by

$$U(2)_+ = \rho_+(Sp(1))\rho_-(U(1)), \qquad U(2)_- = \rho_+(U(1))\rho_-(Sp(1)), \quad (3)$$

where $U(1) \subset Sp(1)$ is the set of unit complex numbers. Since the kernel of $\rho_+\rho_- : Sp(1) \times Sp(1) \to G_2$ is $\pm(1, 1)$, we see that $U(2)_\pm$ are isomorphic to the unitary group $U(2)$.

Further, $SO(4)$ and $SU(3)$ are embedded in G_2 as

$$SO(4) = \rho_+(Sp(1))\rho_-(Sp(1)), \qquad SU(3) = \{g \in G_2 \mid g(i) = i\}. \quad (4)$$

Then we obtain the sequence $Sp(1)_\pm \subset U(2)_\pm \subset SO(4)$ and $U(2)_- \subset SU(3)$. In this way, we obtain the diagram in Introduction.

3. Canonical form

Let G be a compact simple Lie group, and $\mathfrak{g}$ be its Lie algebra. Then there is an adjoint-invariant negative definite symmetric bilinear form $(\ ,\ )$ on $\mathfrak{g}$ uniquely up to multiplication of positive constants. We can extend $(\ ,\ )$ to the complex bilinear form on $\mathfrak{g}^{\mathbb{C}} = \mathfrak{g} \otimes_{\mathbb{R}} \mathbb{C}$.

If we fix such a bilinear form $(\ ,\ )$ on $\mathfrak{g}^{\mathbb{C}}$, we can identify $x \in \mathfrak{g}^{\mathbb{C}}$ with $\check{x} \in (\mathfrak{g}^{\mathbb{C}})^*$ so that $\check{x}(y) = (x, y)$ for every $y \in \mathfrak{g}^{\mathbb{C}}$. Then we can define the dual bilinear form $(\ ,\ )$ on $(\mathfrak{g}^{\mathbb{C}})^*$ by $(\check{x}, \check{y}) = (x, y)$. The bilinear form $(\ ,\ )$ is called *canonical* if

$$(\alpha, \alpha) = 2 \tag{5}$$

for any maximal root α of $\mathfrak{g}$. The canonical bilinear form exists uniquely, and we denote it as $(\ ,\ )_{\mathfrak{g}}$. (See [7] for the detail.)

Now we determine the canonical form of $Sp(1)$ and G_2. First we notice $Sp(1) \simeq SU(2)$. The Lie algebra $\mathfrak{sp}(1) \simeq \mathrm{Im}\mathbb{H}$ is naturally identified with

$$\mathfrak{su}(2) = \left\{ X \in \mathfrak{gl}(2, \mathbb{R}) \mid {}^t\overline{X} = -X, \mathrm{tr}X = 0 \right\}$$

by the correspondence

$$xi + yj + zk \quad \longleftrightarrow \quad \begin{pmatrix} xi & -y + iz \\ y + iz & -xi \end{pmatrix}. \tag{6}$$

Let T be the set of diagonal matrices in

$$SU(2) = \left\{ g \in GL(2, \mathbb{R}) \mid {}^t\overline{g} = g^{-1}, \det g = 1 \right\},$$

then $T \simeq U(1)$ is a maximal torus of $SU(2)$. Let $\mathfrak{a}$ be the corresponding maximal abelian subalgebra of $\mathfrak{su}(2)$. Then $\mathfrak{a}$ and its complexification $\mathfrak{a}^{\mathbb{C}}$ are given as

$$\mathfrak{a} = \mathbb{R}\,X_0, \qquad \mathfrak{a}^{\mathbb{C}} = \mathbb{C}\,X_0 \qquad \text{where } X_0 = \begin{pmatrix} i & 0 \\ 0 & -i \end{pmatrix}.$$

Let us define $\varepsilon \in (\mathfrak{a}^{\mathbb{C}})^*$ by $\varepsilon(iX_0) = 1$. Then the root system is given by $\Delta_{\mathfrak{su}(2)} = \{2\varepsilon, -2\varepsilon\}$.

Proposition 3.1. *The canonical form of $SU(2)$ is given by*

$$(X, Y)_{\mathfrak{su}(2)} = \mathrm{tr}XY \qquad \text{for any } X, Y \in \mathfrak{su}(2). \tag{7}$$

Proof. Let us define a bilinear form $(\ ,\ )$ on $\mathfrak{su}(2)$ by $(X, Y) = \mathrm{tr}XY$. Then this is adjoint-invariant, negative definite, and symmetric. With respect to $(\ ,\ )$, we obtain $2\varepsilon = i\check{X}_0$. Hence for the maximal root 2ε, we obtain

$$(2\varepsilon, 2\varepsilon) = (iX_0, iX_0) = -\mathrm{tr}X_0^2 = 2. \tag{8}$$

Thus $(\,,\,)$ is the canonical form.   □

Remark 3.1. Let $\langle \alpha, \beta \rangle = \mathrm{Re}(\alpha\bar{\beta})$ be the standard inner product on $\mathfrak{sp}(1) \simeq \mathrm{Im}\mathbb{H}$. Then, by the identification (6), we have $(\alpha, \beta)_{\mathfrak{sp}(1)} = -2\langle \alpha, \beta \rangle$.

Next we deal with the Lie group $G_2 = \mathrm{Aut}\mathbb{O}$. We set

$$(e_0, e_1, e_2, e_3, e_4, e_5, e_6, e_7) = (1, i, j, k, \epsilon, i\epsilon, j\epsilon, k\epsilon). \tag{9}$$

Then $G_2 = \mathrm{Aut}\mathbb{O}$ and $\mathfrak{g}_2 \subset \mathrm{End}\mathbb{O}$ are represented as 7 times 7 real matrices with respect to the basis $(e_1, \cdots, e_7)$. Let E_{ij} be the matrix unit, that is

$$E_{ij}(e_k) = \begin{cases} e_i & \text{if } j = k, \\ 0 & \text{otherwise.} \end{cases} \tag{10}$$

We write $G_{ij} = E_{ij} - E_{ji}$. It is known (see [6] for example) that the Lie algebra $\mathfrak{g}_2$ is spanned by the following elements $(a, b, c \in \mathbb{R}, a + b + c = 0)$

$$\begin{aligned} &aG_{23} + bG_{45} + cG_{76}, \\ &aG_{31} + bG_{46} + cG_{57}, \quad aG_{12} + bG_{47} + cG_{65}, \\ &aG_{51} + bG_{73} + cG_{62}, \quad aG_{14} + bG_{72} + cG_{36}, \\ &aG_{71} + bG_{42} + cG_{35}, \quad aG_{61} + bG_{34} + cG_{25}. \end{aligned} \tag{11}$$

Furthermore, we can choose a maximal abelian subalgebra $\mathfrak{a} \subset \mathfrak{g}_2$ by

$$\mathfrak{a} = \left\{ aG_{23} + bG_{45} + cG_{76} \mid a, b, c \in \mathbb{R}, a + b + c = 0 \right\}. \tag{12}$$

Its complexification is

$$\mathfrak{a}^{\mathbb{C}} = \left\{ aG_{23} + bG_{45} + cG_{76} \mid a, b, c \in \mathbb{C}, a + b + c = 0 \right\}. \tag{13}$$

Let us define $\varepsilon_1, \varepsilon_2, \varepsilon_3 \in (\mathfrak{a}^{\mathbb{C}})^*$ so that, for $H = \sqrt{-1}(aG_{23} + bG_{45} + cG_{76})$, they satisfy

$$\varepsilon_1(H) = a, \qquad \varepsilon_2(H) = b, \qquad \varepsilon_3(H) = c. \tag{14}$$

Then we have $\varepsilon_1 + \varepsilon_2 + \varepsilon_3 = 0$, and the root system is given by

$$\Delta_{\mathfrak{g}_2} = \{\pm\varepsilon_1, \pm\varepsilon_2, \pm\varepsilon_3, \pm(\varepsilon_1 - \varepsilon_2), \pm(\varepsilon_2 - \varepsilon_3), \pm(\varepsilon_3 - \varepsilon_1)\}. \tag{15}$$

The maximal roots are $\{\pm(\varepsilon_1 - \varepsilon_2), \pm(\varepsilon_2 - \varepsilon_3), \pm(\varepsilon_3 - \varepsilon_1)\}$.

Proposition 3.2. *The canonical form of* G_2 *is given by*

$$(X, Y)_{\mathfrak{g}_2} = \frac{1}{2}\mathrm{tr}XY \qquad \text{for any } X, Y \in \mathfrak{g}_2. \tag{16}$$

Proof. Let us define a bilinear form $(\,,\,)$ on $\mathfrak{g}_2$ by $(X,Y) = \frac{1}{2}\mathrm{tr}XY$. Then this is adjoint-invariant, negative definite, and symmetric. We take a basis $\{u_1, u_2\}$ of $\mathfrak{a}^{\mathbb{C}}$ by

$$u_1 = \sqrt{\frac{-1}{2}}\,(G_{45} - G_{76}), \qquad u_2 = \sqrt{\frac{-1}{6}}\,(2G_{23} - G_{45} - G_{76}). \qquad (17)$$

Then $\{u_1, u_2\}$ is an orthonormal basis, that is,

$$(u_i, u_j) = \begin{cases} 1 & \text{if } i = j, \\ 0 & \text{if } i \neq j. \end{cases} \qquad (18)$$

By (14), we have

$$\varepsilon_1(u_1) = 0, \quad \varepsilon_1(u_2) = \frac{2}{\sqrt{6}}, \quad \varepsilon_2(u_1) = \frac{1}{\sqrt{2}}, \quad \varepsilon_2(u_2) = \frac{-1}{\sqrt{6}}. \qquad (19)$$

Hence

$$\varepsilon_1 = \frac{2}{\sqrt{6}}\,\breve{u}_2, \qquad \varepsilon_2 = \frac{1}{\sqrt{2}}\,\breve{u}_1 - \frac{1}{\sqrt{6}}\,\breve{u}_2, \qquad (20)$$

where $\{\breve{u}_1, \breve{u}_2\}$ is the dual orthonormal basis of $\{u_1, u_2\}$. Then we obtain $(\varepsilon_1, \varepsilon_1) = (\varepsilon_2, \varepsilon_2) = \frac{2}{3}$ and $(\varepsilon_1, \varepsilon_2) = -\frac{1}{3}$. For the maximal root $\alpha = \varepsilon_1 - \varepsilon_2$, we have $(\alpha, \alpha) = 2$. Hence $(\,,\,)$ is the canonical form of G_2. $\square$

4. Index

Here, we deal with the *index* (Dynkin index) (see [7] 3.10). Let $\rho : H \to G$ be a homomorphism of compact simple Lie groups. Let $\rho_* : \mathfrak{h} \to \mathfrak{g}$ be the corresponding Lie algebra homomorphism. Then by the uniqueness of the canonical form, there exists a constant c such that,

$$(\rho_*(x), \rho_*(x))_{\mathfrak{g}} = c\,(x, x)_{\mathfrak{h}}, \qquad (21)$$

for any $x \in \mathfrak{h}$. This constant c is called the *index* of ρ.

Proposition 4.1. *The index of $\rho_+ : Sp(1)_+ \to G_2$ is 1, while the index of $\rho_- : Sp(1)_- \to G_2$ is 3.*

Proof. The differentials $(\rho_\pm)_* : \mathfrak{sp}(1) = \mathrm{Im}\mathbb{H} \to \mathfrak{g}_2$ are given by

$$(\rho_+)_*(X)(a + b\epsilon) = (Xb)\epsilon, \qquad (\rho_-)_*(X)(a + b\epsilon) = [X,a] - (bX)\epsilon \qquad (22)$$

for $X \in \mathfrak{sp}(1) = \mathrm{Im}\mathbb{H}$. Then we obtain

$$(\rho_+)_*(i) = G_{45} - G_{76}, \qquad (\rho_-)_*(i) = 2G_{23} - G_{45} - G_{76}. \qquad (23)$$

By Propositions 3.1 and 3.2, the index of ρ_+ is

$$\frac{\big((\rho_+)_*(i),(\rho_+)_*(i)\big)_{\mathfrak{g}_2}}{(i,i)_{\mathfrak{sp}(1)}} = \frac{-2}{-2} = 1. \tag{24}$$

Similarly, the index of ρ_- is

$$\frac{\big((\rho_-)_*(i),(\rho_-)_*(i)\big)_{\mathfrak{g}_2}}{(i,i)_{\mathfrak{sp}(1)}} = \frac{-6}{-2} = 3. \tag{25}$$

$\square$

5. Main theorem

We recall the following facts on the homotopy theory.

Theorem 5.1. *The following holds:*

(1) *For any compact simple Lie group G, we have $\pi_2(G) = 0$ and $\pi_3(G) \simeq \mathbb{Z}$.*

(2) *Let $\rho : H \to G$ be a homomorphism of compact simple Lie groups of index c. Then the induced map $\rho_* : \pi_3(H) \to \pi_3(G)$ is the multiplication with c for a suitable choice of generators.*

We give a direct proof of the first statement of Theorem 5.1 for $G = Sp(1), U(2)$ and G_2 in Proposition 6.1. For the general case, see [1], [7] and [11].

Theorem 5.2. *The third homotopy groups of the homogeneous spaces $G_2/Sp(1)_\pm$ are $\pi_3(G_2/Sp(1)_+) = 0$ and $\pi_3(G_2/Sp(1)_-) \simeq \mathbb{Z}/3\mathbb{Z}$. Consequently $G_2/Sp(1)_+$ and $G_2/Sp(1)_-$ are not homeomorphic to each other.*

Proof. We consider the fibration

$$Sp(1)_+ \xrightarrow{\rho_+} G_2 \longrightarrow G_2/Sp(1)_+. \tag{26}$$

Then the induced exact sequence is

$$\pi_3(Sp(1)_+) \xrightarrow{(\rho_+)_*} \pi_3(G_2) \longrightarrow \pi_3(G_2/Sp(1)_+) \longrightarrow \pi_2(Sp(1)_+). \tag{27}$$

Recall that the index of ρ_+ is 1 (Proposition 4.1). By Theorem 5.1, the exact sequence (27) is written as

$$\mathbb{Z} \xrightarrow{\simeq} \mathbb{Z} \longrightarrow \pi_3(G_2/Sp(1)_+) \longrightarrow 0. \tag{28}$$

Hence we get $\pi_3(G_2/Sp(1)_+) = 0$. Similarly, the fibration

$$Sp(1)_- \xrightarrow{\rho_-} G_2 \longrightarrow G_2/Sp(1)_-$$

induces an exact sequence

$$\pi_3(Sp(1)_-) \xrightarrow{(\rho_-)_*} \pi_3(G_2) \longrightarrow \pi_3(G_2/Sp(1)_-) \longrightarrow \pi_2(Sp(1)_-)$$

which is rewriten as

$$\mathbb{Z} \longrightarrow \mathbb{Z} \longrightarrow \pi_3(G_2/Sp(1)_-) \longrightarrow 0.$$

Here $(\rho_-)_*$ corresponds to the map $j \mapsto 3j$ by Theorem 5.1. Hence, we obtain $\pi_3(G_2/Sp(1)_-) \simeq \mathbb{Z}/3\mathbb{Z}$.   □

Lemma 5.1. *For the inclusion map $i_\pm : Sp(1)_\pm \to U(2)_\pm$, the induced map $(i_\pm)_* : \pi_n(Sp(1)_\pm) \to \pi_n(U(2)_\pm)$ is an isomorphism for $n > 1$.*

Proof. Since the kernel of $\rho_+\rho_- : Sp(1) \times Sp(1) \to G_2$ is $\pm(1,1)$, we have

$$U(2)_\pm/Sp(1)_\pm \simeq U(1)/\{\pm 1\} \simeq S^1. \qquad (29)$$

Hence we obtain a fibration

$$Sp(1)_\pm \xrightarrow{i_\pm} U(2)_\pm \longrightarrow S^1, \qquad (30)$$

and an induced exact sequence

$$\pi_{n+1}(S^1) \longrightarrow \pi_n(Sp(1)_\pm) \xrightarrow{(i_\pm)_*} \pi_n(U(n)_\pm) \longrightarrow \pi_n(S^1) \qquad (31)$$

for $n \geq 1$. Since $\pi_n(S^1) = 0$ for $n > 1$, the statement follows.   □

Theorem 5.3. *The third homotopy groups of the homogeneous spaces $G_2/U(2)_\pm$ are $\pi_3(G_2/U(2)_+) = 0$ and $\pi_3(G_2/U(2)_-) \simeq \mathbb{Z}/3\mathbb{Z}$. Consequently $G_2/U(2)_+$ and $G_2/U(2)_-$ are not homeomorphic to each other.*

Proof. The sequence of inclusion maps $Sp(1)_\pm \xrightarrow{i_\pm} U(2)_\pm \xrightarrow{j_\pm} G_2$ induces the homomorphisms

$$\pi_3(Sp(1)_\pm) \xrightarrow{(i_\pm)_*} \pi_3(U(2)_\pm) \xrightarrow{(j_\pm)_*} \pi_3(G_2). \qquad (32)$$

Since $(i_+)_*$ is an isomorphism (Lemma 5.1) and the index of $\rho_+ = j_+ \circ i_+$ is 1 (Proposition 4.1), we see that the image of a generator of $\pi_3(U(2)) \simeq \mathbb{Z}$ is a generator of $\pi_3(G_2)$. Then, by Theorem 5.2, we obtain $\pi_3(G_2/U(2)_+) = 0$. Similarly we can check $\pi_3(G_2/U(2)_-) \simeq \mathbb{Z}/3\mathbb{Z}$.   □

6. Calculation of the homotopy groups

Here we give a direct proof of the following

Proposition 6.1. *For $G = Sp(1), U(2)$, and G_2, we obtain*

$$\pi_2(G) = 0, \qquad \pi_3(G) \simeq \mathbb{Z}. \tag{33}$$

Proof. As is well known (see [12],15.9), for the m-dimensional sphere S^m, we have

$$\pi_n(S^m) \simeq \begin{cases} 0 & \text{if} \quad (n < m) \quad \text{or} \quad (n > m = 1), \\ \mathbb{Z} & \text{if} \quad n = m. \end{cases} \tag{34}$$

Since $Sp(1) \simeq SU(2) \sim S^3$, we obtain $\pi_2(Sp(1)) = 0$ and $\pi_3(Sp(1)) \simeq \mathbb{Z}$. Next we consider the fibration

$$SU(2) \simeq S^3 \longrightarrow U(2) \xrightarrow{\det} U(1) \simeq S^1. \tag{35}$$

By the induced long exact sequence, we see

$$\pi_2(U(2)) = 0, \qquad \pi_3(U(2)) \simeq \pi_3(S^3) \simeq \mathbb{Z}. \tag{36}$$

The natural map $\mathbb{C}^3 \supset S^5 \to \mathbb{C}P^2$ gives an S^1-fibration

$$S^1 \longrightarrow S^5 \longrightarrow \mathbb{C}P^2, \tag{37}$$

of which the long exact sequence gives $\pi_3(\mathbb{C}P^2) = \pi_4(\mathbb{C}P^2) = 0$. We note $\mathbb{C}P^2 \simeq SU(3)/U(2)$, or equivalently we have a fibration $U(2) \to SU(3) \to \mathbb{C}P^2$. By the induced long exact sequence, we have

$$\pi_3(SU(3)) \simeq \pi_3(U(2)). \tag{38}$$

Finally, we note $S^6 \simeq G_2/SU(3)$, or equivalently we have a fibration $SU(3) \to G_2 \to S^6$. Therefore, we obtain

$$\pi_3(G_2) \simeq \pi_3(SU(3)). \tag{39}$$

Thus we get $\pi_2(G_2) = 0$ and $\pi_3(G_2) \simeq \mathbb{Z}$. □

Acknowledgement

The author would like to thank Hideya Hashimoto, Katsuya Mashimo and Misa Ohashi for helpful conversations and for providing many useful references.

References

[1] R. Bott, An application of the Morse theory to the topology of Lie groups, *Bull. Soc. Math. France*, **84**, 251-281 (1956).

[2] L. Besse, *Einstein manifolds*, Springer (1987).

[3] C. Boyer, K. Galicki, *Sasakian Geometry* Oxford Mathematical Monographs, Oxford Science Publications (2008).

[4] R. L. Bryant, *Submanifolds and special structures on the octonions*, J. Diff. Geom. **17** (1982), 185-232.

[5] R. Harvey, H. B. Lawson, Calibrated geometries, *Acta Math.* **148**, 47-157 (1982).

[6] H. Hashimoto, T. Koda, K. Mashimo, K. Sekigawa Extrinsic homogeneous almost hermitian 6-dimensional submanifolds in the octonions, *Kodai Math. J.*, **30**, 297-321 (2007).

[7] A. L. Onishchik, *Topology of transitive transformation groups*, Johann Ambrosius Barth Verlag GmbH, Leipzig, (1994).

[8] S. Salamon, Quaternionic Kähler manifolds, *Invent. Math.*, **67**, 143-171 (1982).

[9] S. Salamon, *Riemannian geometry and holonomy groups*, Pitman research notes in mathematics series, (1989).

[10] R. Penrose, Nonlinear gravitons and curved twistor theory, *Gen. Rel. Grav.*, **7**, 31-52 (1976).

[11] A. Presley, G. Segal, *Roop groups*, Oxford Mathematical Monograph, Oxford Science Publications, (1986).

[12] N. E. Steenrod, *Topology of fibre bundles* Princeton University Press, Princeton, N. J. (1951).

Received July 10, 2017
Revised   July 24, 2017

SEQUENTIAL STRUCTURE OF STATISTICAL
MANIFOLDS AND ITS DIVERGENCE GEOMETRY

Hiroshi MATSUZOE

Department of Computer Science and Engineering,
Graduate School of Engineering, Nagoya Institute of Technology,
Nagoya, Aichi 466-8555 Japan
E-mail: matsuzoe@nitech.ac.jp

We give a survey on sequential structures of escort distributions and sequential structures of statistical manifolds. For a deformed exponential family, the notion of escort distribution has been introduced in nonextensive statistical physics, and an expectation with respect to its escort distribution has been discussed. The author introduced a sequential structure of escort distributions, and discussed statistical manifold structures derived from sequential escort distributions. In this paper, after reviewing basic facts on information geometry, we study sequential structures of statistical manifolds. A divergence function which is a distance-like function on a statistical manifold plays an important role in this framework.

Keywords: Escort distribution; escort expectations; statistical manifold; information geometry; Tsallis statistics; divergence geometry.

1. Introduction

An exponential family plays an important role in the theory of statistical inferences. If we can obtain a large number of i.i.d.-observations, an exponential family works effectively. When random variables have correlations, or when we obtain a few number of observations, non-exponential type statistical models are also important. A deformed exponential family is a generalization of exponential family, and it is introduced in nonexpensive statistical physics. Since a deformed exponential family includes heavily tailed probability distributions, it is useful for mathematical modelings of non-standard phenomenon [22].

However, since a generalized exponential family includes heavily tailed distributions, expectations and variances do not exist in general. Therefore, the notion of escort distributions was introduced [4, 17] to give a suitable weight for a probability distribution, and an expectation with respect to the escort distribution was studied. The author recently showed that a deformed exponential family naturally has a sequential structure of escort distributions. Consequently, a sequential structure of statistical manifolds is obtained from the sequence of escort distributions. Since several kinds of statistical manifolds which are known in information geometry are included in this sequence, the sequential structure is quite natural for the geometry of probability distributions of non-exponential type.

In this paper, after reviewing preliminaries on information geometry [1, 2], we survey sequential structures of escort distributions and statistical manifold structures derived from sequential escort distributions [12, 15]. The geometry induced from divergence functions is also important in this framework [5, 11].

2. Geometry of statistical models

Throughout this paper, we assume that all objects are smooth. We begin with reviewing of geometry of statistical models.

2.1. Statistical models

Let $(\Omega, \mathcal{F})$ be a measurable space. Suppose that $p(x; \xi)$ is a probability density function on Ω parametrized by $\xi = (\xi^1, \ldots, \xi^n) \in \Xi \subset \mathbf{R}^n$, where Ξ is an open domain on $\mathbf{R}^n$. We say that a set S is a *statistical model* or a *parametric model* if

$$S = \left\{ p(x; \xi) \,\middle|\, \int_\Omega p(x; \xi) dx = 1, \; p(x; \xi) > 0, \; \xi \in \Xi \subset \mathbf{R} \right\}.$$

We regard that S is a manifold with a coordinate system $(\xi^1, \ldots, \xi^n)$.

We denote by $\tilde{\partial}_i = \partial/\partial \xi^i$ and $l_\xi - l_\xi(x) = \ln p(x; \xi)$. We define a $(0, 2)$-tensor $g^F(\xi) = (g_{ij}^F(\xi)) = (g^F((\tilde{\partial}_i)_\xi, (\tilde{\partial}_j)_\xi))$ by

$$g_{ij}^F(\xi) = E_p[\tilde{\partial}_i l_\xi \tilde{\partial}_j l_\xi]$$

$$= \int_\Omega \left(\tilde{\partial}_i \ln p(x; \xi) \right) \left(\tilde{\partial}_j \ln p(x; \xi) \right) p(x; \xi) \, dx, \tag{1}$$

$$= \int_\Omega \left(\tilde{\partial}_i \ln p(x; \xi) \right) \left(\tilde{\partial}_j p(x; \xi) \right) \, dx, \tag{2}$$

where $E_p[f(x)]$ is the expectation of $f(x)$ with respect to $p(x; \xi)$. The tensor $g^F(\xi)$ is called a *Fisher information matrix*, and it is nonnegative definite in general. By assuming that $g^F(\xi)$ is positive definite, we can define a Riemannian metric g^F on S, which is called a *Fisher metric*. The following proposition is known in information geometry (Chapter 2 in [2]).

Proposition 2.1. *For a statistical model S, the following conditions are equivalent.*

(1) $g^F(\xi) = (g^F_{ij}(\xi))$ *is positive definite.*

(2) $\{\tilde{\partial}_1 l_\xi, \tilde{\partial}_2 l_\xi, \ldots, \tilde{\partial}_n l_\xi\}$ *are linearly independent as functions on* Ω.

(3) $\{\tilde{\partial}_1 p_\xi, \tilde{\partial}_2 p_\xi, \ldots, \tilde{\partial}_n p_\xi\}$ *are linearly independent as functions on* Ω.

From Proposition 2.1, functions $\tilde{\partial}_i \ln p(x;\xi)$ and $\tilde{\partial}_i p(x;\xi)$ are regarded as tangent vectors on $T_{p(x;\xi)}S$, intuitively. We call $\tilde{\partial}_i \ln p(x;\xi)$ an *exponential representation* (or *e-representation*) of $p(x;\xi)$ and $\tilde{\partial}_i p(x;\xi)$ a *mixture representation* (or *m-representation*), respectively.

Let us define an affine connection for S. We define a totally symmetric $(0,3)$-tensor field $C^F = (C^F_{ijk})$ on S, called a *cubic form* or an *Amari-Chentsov tensor field*, by

$$C^F_{ijk}(\xi) = E_p[\tilde{\partial}_i l_\xi \tilde{\partial}_j l_\xi \tilde{\partial}_k l_\xi]$$
$$= \int_\Omega \left(\tilde{\partial}_i \ln p(x;\xi)\right)\left(\tilde{\partial}_j \ln p(x;\xi)\right)\left(\tilde{\partial}_k \ln p(x;\xi)\right) p(x;\xi)\, dx. \quad (3)$$

Using this tensor field, for a fixed $\alpha \in \mathbf{R}$, we define an α-connection $\nabla^{(\alpha)}$ on S by

$$g^F(\nabla^{(\alpha)}_X Y, Z) := g^F(\nabla^{(0)}_X Y, Z) - \frac{\alpha}{2} C^F(X, Y, Z), \quad (4)$$

where X, Y and Z are arbitrary vector fields on S. We denoted by $\nabla^{(0)}$ the Levi-Civita connection with respect to the Fisher metric g^F.

Two connections $\nabla^{(\alpha)}$ and $\nabla^{(-\alpha)}$ are *mutually dual* with respect to g^F, that is, the following equation holds.

$$X g^F(Y, Z) = g^F(\nabla^{(\alpha)}_X Y, Z) + g^F(Y, \nabla^{(-\alpha)}_X Z).$$

It is easy to check that $\nabla^{(\alpha)}$ is torsion-free, and that the covariant derivative $\nabla^{(\alpha)} g^F$ is totally symmetric. These imply that the triplet $(S, \nabla^{(\alpha)}, g^F)$ (or (S, g^F, C^F)) is a *statistical manifold* [8, 10].

When $\alpha = \pm 1$, we have the following expressions:

$$\Gamma^{(e)}_{ij,k}(\xi) := g^F(\nabla^{(1)}_{\tilde{\partial}_i} \tilde{\partial}_j, \tilde{\partial}_k) = E_p\left[(\tilde{\partial}_i \tilde{\partial}_j \ln p(x;\xi))(\tilde{\partial}_k p(x;\xi))\right], \quad (5)$$

$$\Gamma^{(m)}_{ik,j}(\xi) := g^F(\tilde{\partial}_j, \nabla^{(-1)}_{\tilde{\partial}_i} \tilde{\partial}_k) = E_p\left[(\tilde{\partial}_j \ln p(x;\xi))(\tilde{\partial}_i \tilde{\partial}_k p(x;\xi))\right]. \quad (6)$$

The connection $\nabla^{(e)} := \nabla^{(1)}$ is called an *exponential connection* and $\nabla^{(m)} := \nabla^{(-1)}$ is called a *mixture connection*.

2.2. Exponential families

We suppose that a statistical model S_e has the following expression

$$S_e = \left\{ p(x;\theta) \ \middle| \ p(x;\theta) = \exp\left[\sum_{i=1}^{n} \theta^i F_i(x) - \psi(\theta) \right], \ \theta \in \Theta \subset \mathbf{R}^n \right\}, \quad (7)$$

where $F_1(x), \ldots, F_n(x)$ are functions on a total sample space Ω, $\theta = (\theta^1, \ldots, \theta^n) \in \Theta \subset \mathbf{R}^n$ is a parameter, and $\psi(\theta)$ is the normalization term with respect to the parameter θ. We call the statistical model S_e an *exponential family*. We assume that the parameter space Θ is an open domain in $\mathbf{R}^n$.

From the definition of exponential family and Equation (5), we find that $\Gamma_{ij,k}^{(e)}$ always vanishes. Therefore, the exponential connection $\nabla^{(e)}$ is flat, and θ is an affine coordinate system on S_e. We call the coordinate system $\theta = (\theta^1, \ldots \theta^n)$ a *natural coordinate system*. In this case, the mixture connection $\nabla^{(m)}$ is also flat. Then the quadruplet $(S_e, g^F, \nabla^{(e)}, \nabla^{(m)})$ is a *dually flat space*, and the triplet $(g^F, \nabla^{(e)}, \nabla^{(m)})$ is a *dually flat structure* on S_e. It is known that there exists a $\nabla^{(m)}$-affine coordinate system $\eta = (\eta_1, \ldots, \eta_n)$ such that

$$g^F\left(\partial_i, \partial^j\right) = \delta_i^j,$$

where $\partial_i = \partial/\partial\theta^i$ and $\partial^j = \partial/\partial\eta_j$, The coordinate system (η_i) is called the *dual coordinate system* of θ with respect to g^F. It is also called an *expectation coordinate system*, since each η_i is given by

$$\eta_i := E_p[F_i(x)], \quad (i = 1, 2, \ldots, n).$$

Proposition 2.2. *Let $(g^F, \nabla^{(e)}, \nabla^{(m)})$ be a dually flat structure on S_e. Suppose that (θ^i) is a natural coordinate system, and (η_i) is its dual coordinate system. Then there exists a function ϕ on S_e such that*

$$\frac{\partial\psi}{\partial\theta^i} = \eta_i, \quad \frac{\partial\phi}{\partial\eta_i} = \theta^i, \quad \psi(p) + \phi(p) - \sum_{i=1}^{n} \theta^i(p)\eta_i(p) = 0, \quad (\forall p \in S_e).$$

In addition, two functions ψ and ϕ are potentials of the Fisher metric, that is,

$$g_{ij}^F = \frac{\partial^2\psi}{\partial\theta^i\partial\theta^j}, \quad g^{F\,ij} = \frac{\partial^2\phi}{\partial\eta_i\partial\eta_j},$$

where $(g^{F\,ij})$ is its inverse matrix of the Fisher metric $(g_{ij}^F) = (g^F(\partial_i.\partial_j))$. The cubic form C for $(S_e, g^F, \nabla^{(e)}, \nabla^{(m)})$ is given by

$$C_{ijk} = \frac{\partial^3\psi}{\partial\theta^i\partial\theta^j\partial\theta^k}.$$

We remark that we can choose ϕ as the (negative) *entropy functional* on S_e defined by

$$\phi(p) := E_p[\ln p(x;\theta)].$$

Using those two functions ψ and ϕ, we define a *canonical divergence* of $(S_e, g^F, \nabla^{(e)}, \nabla^{(m)})$ by

$$D(p,r) := \psi(\theta(p)) + \phi(\eta(r)) - \sum_{i=1}^{n} \theta^i(p)\eta_i(r),$$

where $p := p(x;\theta)$ and $r := p(x;\theta')$ are arbitrary points in S_e. We remark that the definition is independent of a choice of dual affine coordinate systems θ and η. Since the canonical divergence D is not a symmetric function, it is not a distance on S_e. However D measures a dissimilarity of two points between p and r.

On the other hand, a *Kullback-Leibler divergence* D_{KL} or a *relative entropy* on S_e is defined by

$$D_{KL}(p,r) := E_p[\ln p(x) - \ln r(x)] = \int_\Omega p(x;\theta)\frac{p(x;\theta)}{p(x;\theta')}dx.$$

Since S_e is an exponential family, the Kullback-Leibler divergence can be given by

$$D_{KL}(p,r) = E_p\left[\left(\sum_{i=1}^{n}\theta^i F_i(x) - \psi(\theta)\right) - \left(\sum_{i=1}^{n}(\theta')^i F_i(x) - \psi(\theta')\right)\right]$$

$$= \left(\sum_{i=1}^{n}\theta^i\eta_i - \psi(\theta)\right) - \left(\sum_{i=1}^{n}(\theta')^i\eta_i - \psi(\theta')\right)$$

$$= \psi(\theta') + \phi(\theta) - \sum_{i=1}^{n}(\theta')^i\eta_i$$

$$= D(r,p) \tag{8}$$

Therefore, the Kullback-Leibler divergence $D_{KL}(p,r)$ coincides with the canonical divergence $D(r,p)$.

2.3. Geometric interpretations of statistical inference

Let $S = \{p(x;\xi) \mid \xi \in \Xi\}$ be a statistical model, and let $x_1,\ldots,x_N$ be N-independent observations generated from a probability density function $p(x;\xi) \in S$. We define a *likelihood function* $L(\xi)$ by

$$L(\xi) = p(x_1;\xi)p(x_2;\xi)\cdots p(x_N;\xi). \tag{9}$$

The parameter $\hat{\theta}$ which attains the maximum of this likelihood function L is the *maximum likelihood estimator* (MLE), that is,

$$\hat{\theta} := \arg\max_{\xi \in \Xi} L(\xi).$$

Let us consider geometry of statistical inferences. Let S_e be an exponential family, and M be a submanifold in S_e. In statistics, M is called a *curved exponential family* in S_e. Suppose that $x_1, \ldots, x_N$ are N-independent observations generated from $p(x; u) = p(x; \theta(u)) \in M$.

By taking a logarithm, the log-likelihood function is written by

$$\ln L(u) = \sum_{j=1}^{N} \ln p(x_j; u)$$

$$= \sum_{j=1}^{N} \left\{ \sum_{i=1}^{n} \theta^i(u) F_i(x_j) - \psi(\theta(u)) \right\}$$

$$= \sum_{i=1}^{n} \theta^i(u) \sum_{j=1}^{N} F_i(x_j) - N\psi(\theta(u)).$$

Since an MLE attains the maximum of the likelihood function L, we obtain a log-likelihood equation

$$\partial_i \ln L(u) = \sum_{j=1}^{N} F_i(x_j) - N\partial_i\psi(\theta(u)) = 0.$$

Then the MLE for S_e is given by

$$\hat{\eta}_i = \frac{1}{N} \sum_{j=1}^{N} F_i(x_j).$$

From Equation (8), the Kullback-Leibler divergence can be written by

$$D_{KL}(p(\hat{\eta}), p(\theta(u))) = D(p(\theta(u)), p(\hat{\eta}))$$

$$= \psi(\theta(u)) + \phi(\hat{\eta}) - \sum_{i=1}^{n} \theta^i(u)\hat{\eta}_i$$

$$= \phi(\hat{\eta}) - \frac{1}{N} \ln L(u).$$

Therefore, the likelihood attains the maximum if and only if the Kullback-Leibler divergence from $p(\hat{\eta}) \in S_e$ to the curved exponential family M attains the minimum. This implies that the MLE is obtained by the Kullback-Leiber divergence projection from $\hat{\eta}$ to the model distribution M.

3. Deformed exponential families

In this section, we review preliminaries of deformed exponential families.
Let $\mathbf{R}_{++}$ be the set of all positive real numbers, i.e. $\mathbf{R}_{++} := \{x \in \mathbf{R} \mid x > 0\}$. Let χ be a monotone increasing function from $\mathbf{R}_{++}$ to $\mathbf{R}_{++}$. We define a χ-*logarithm function* or a *deformed logarithm function* [17, 18] by

$$\ln_\chi s := \int_1^s \frac{1}{\chi(t)} dt.$$

We call χ a *deformation function* of this generalized logarithm. If χ is an identity function, that is, $\chi(t) = t$, then the natural logarithm $\ln s$ is recovered.

The inverse of $\ln_\chi s$ is given by

$$\exp_\chi t := 1 + \int_0^t u(s) ds,$$

where the function $u(s)$ satisfies $u(\ln_\chi s) = \chi(s)$. The function $\exp_\chi t$ is called a χ-*exponential function* or a *deformed exponential function*.

Let us give examples of deformed exponential functions. If a deformation functions χ is a power function, that is, $\chi(t) = t^q$ $(q > 0, q \neq 1)$, the deformed logarithm and the deformed exponential are defined by

$$\ln_q s := \frac{s^{1-q} - 1}{1 - q}, \qquad\qquad (s > 0),$$

$$\exp_q t := \left(1 + (1 - q)t\right)^{1/(1-q)}, \qquad \left(1 + (1 - q)t > 0\right).$$

We call $\ln_q s$ a q-*logarithm function* and $\exp_q t$ a q-*exponential function*. In this case, the function $u_q(s)$ is given by

$$u_q(s) = \left(1 + (1 - q)s\right)^{q/(1-q)} = \{\exp_q s\}^q.$$

If $\chi(t) = 2t/(t^\kappa + t^{-\kappa})$ $(-1 < \kappa < 1, \kappa \neq 0)$, we have

$$\ln_\kappa s := \frac{s^\kappa - s^{-\kappa}}{2\kappa}, \qquad\qquad (s > 0),$$

$$\exp_\kappa t := \left(\kappa t + \sqrt{1 + \kappa^2 t^2}\right)^{1/\kappa}.$$

We call $\ln_\kappa s$ a κ-*logarithm function* and $\exp_\kappa t$ a κ-*exponential function* [6]. In this case, the function $u_\kappa(s)$ is given by

$$u_\kappa(s) = \frac{\left(\kappa s + \sqrt{1 + \kappa^2 s^2}\right)^{1/\kappa}}{\sqrt{1 + \kappa^2 s^2}}.$$

We suppose that a statistical model S_χ has the following expression

$$S_\chi := \left\{ p(x, \theta) \,\middle|\, p(x; \theta) = \exp_\chi \left[\sum_{i=1}^{n} \theta^i F_i(x) - \psi(\theta) \right], \ \theta \in \Theta \subset \mathbf{R}^n \right\}, \quad (10)$$

where $F_1(x), \ldots, F_n(x)$ are functions on a sample space Ω, $\theta = {}^t(\theta^1, \ldots, \theta^n)$ is a parameter, and $\psi(\theta)$ is the normalization defined by $\int_\Omega p(x; \theta) dx = 1$. In this case, $S_\chi i$ is called χ-*exponential family* or a *deformed exponential family* [3, 13, 17]. The function ψ is convex in general. We assume that ψ is strictly convex throughout this paper. Under suitable conditions as in the (standard) exponential family, the statistical model S_χ is regarded as a manifold with a coordinate system $(\theta^1, \ldots, \theta^n)$, which is called a *natural coordinate system*.

If a deformed exponential is a q-exponential $\exp_q t$, the deformed exponential family is said to be a q-*exponential family*, and is denoted by S_q. Similarly, if a deformed exponential is a κ-exponential $\exp_\kappa t$, the family is called a κ-*exponential family*, and is denoted by S_κ.

Intuitively, in the case of an exponential family S_e, random variables and parameters are monotonically embedded into a functional space $L^1(\Omega)$ by use of the exponential function $\exp t$. In the case of a deformed exponential family S_χ, such objects are embedded by use of a deformed exponential $\exp_\chi$ [23].

We give an important example of deformed exponential families.

Example 3.1 (Student's t-distribution or q-normal distribution).
Fix a number q $(1 < q < 1 + 2/d, \ d \in \mathbf{N})$, where d is the dimension of total sample space Ω. By setting $\nu = -d - 2/(1 - q)$, we define a d-*dimensional Student's t-distribution with degree of freedom* ν or a q-*normal distribution* by

$$p_q(x; \mu, \Sigma) := \frac{\Gamma\left(\frac{\nu+d}{2}\right)}{(\pi\nu)^{\frac{d}{2}} \Gamma\left(\frac{\nu}{2}\right) \sqrt{\det(\Sigma)}} \left[1 + \frac{1}{\nu} \, {}^t(x-\mu)\Sigma^{-1}(x-\mu) \right]^{-(\nu+d)/2},$$

where $X = {}^t(X_1, \ldots, X_d)$ is a random vector on $\mathbf{R}^d$, $\mu = {}^t(\mu^1, \ldots, \mu^d)$ is a location vector on $\mathbf{R}^d$ and Σ is a scale matrix on $\mathrm{Sym}^+(d)$. We assume that Σ is invertible. Then, the set of all Student's t-distributions $S_q = \{p_q(x; \mu, \Sigma)\}$ is a q-exponential family.

Set the normalization factors by

$$z_q := \frac{(\pi\nu)^{\frac{d}{2}} \Gamma\left(\frac{\nu}{2}\right) \sqrt{\det(\Sigma)}}{\Gamma\left(\frac{\nu+d}{2}\right)}, \quad \psi(\theta) := \frac{1}{4} \, {}^t\theta \tilde{R}^{-1}\theta - \ln_q \frac{1}{z_q},$$

and the natural coordinates by

$$\tilde{R} = \frac{z_q^{q-1}}{(1-q)d+2}\Sigma^{-1}, \quad \text{and} \quad \theta = 2\tilde{R}\mu. \tag{11}$$

Then we have

$$p_q(x; \mu, \Sigma) = \frac{1}{z_q}\left[1 + \frac{1}{\nu}{}^t(x-\mu)\Sigma^{-1}(x-\mu)\right]^{1/(1-q)}$$

$$= \exp_q\left[\sum_{i=1}^d \theta^i x_i - \sum_{i=1}^d \tilde{R}_{ii}x_i^2 - 2\sum_{i<j}\tilde{R}_{ij}x_i x_j - \psi(\theta)\right].$$

This implies that S_q is a q-exponential family. Since $\theta \in \mathbf{R}^d$ and $\tilde{R} \in \mathrm{Sym}^+(d)$, we find that $\dim S_q = d(d+3)/2$.

For further details on geometry of Student's t-distributions, see [12, 20]. Statistical properties are studied in [7].

4. A sequential structure of expectations

In this section, we consider a sequential structure of expectations. In the case of a deformed exponential family S_χ, the standard expectation may not be natural, and we can consider several generalizations of expectations. Therefore, we introduce a sequential structure of expectations.

Suppose that $S_\chi = \{p_\theta\} = \{p(x;\theta)\}$ is a deformed exponential family. Denote by $\exp^{(n)} x$ the n-th differential of the deformed exponential function. For a given probability density $p \in S_\chi$, the n-th escort distribution $P_{\chi,(n)}(x;\theta)$ is defined by

$$P_{\chi,(n)}(x;\theta) := \exp_\chi^{(n)}(\ln_\chi p_\theta) = \exp_\chi^{(n)}\left(\sum_{i=1}^n \theta^i F_i(x) - \psi(\theta)\right).$$

For example, in the case of q-exponential family S_q, the n-th escort distribution of $p_q(x;\theta) \in S_q$ is given by

$$P_{q,(n)}(x;\theta) = \{q(2q-1)\cdots((n-1)q-(n-2))\}\{p_q(x;\theta)\}^{nq-(n-1)}.$$

In this paper, we assume that each escort distribution $P_{q,(n)}(x;\theta)$ is integrable on Ω. To elucidate regularity conditions for escort distributions is quite an open problem.

The density $P_{\chi,(n)}(x;\theta)$ is not a probability density, but a positive valued density. Therefore, we need a suitable normalization if we have to consider a probability density. The *normalized n-th escort distribution* $P^{esc}_{\chi,(n)}(x;\theta)$ is defined by

$$P^{esc}_{\chi,(n)}(x;\theta) := \frac{1}{Z_{\chi,(n)}(p_\theta)} P_{\chi,(n)}(x;\theta),$$

where

$$Z_{\chi,(n)}(p_\theta) = \int_\Omega P_{\chi,(n)}(x;\theta)dx.$$

The expectations with respect to $P_{\chi,(n)}(x;\theta)$ and $P^{esc}_{\chi,(n)}(x;\theta)$ are called the *n-th escort expectation* and the *normalized n-th escort expectation*, respectively. That is,

$$E_{\chi,(n),p}[f(x)] := \int_\Omega f(x)P_{\chi,(n)}(x;\theta)dx,$$

$$E^{esc}_{\chi,(n),p}[f(x)] := \int_\Omega f(x)P^{esc}_{\chi,(n)}(x;\theta)dx,$$

where $f(x)$ is a function on the total sample space Ω.

For the first escort distributions, we simply call them the *escort distribution*, and the *normalized escort distribution*, respectively, and denote by

$$P_\chi(x;\theta) := P_{\chi,(1)}(x;\theta) := \chi(p_\theta),$$

$$P^{esc}_\chi(x;\theta) := P^{esc}_{\chi,(1)}(x;\theta) := \frac{\chi(p_\theta)}{Z_\chi(p_\theta)},$$

where

$$Z_\chi(p_\theta) := Z_{\chi,(1)}(p_\theta) := \int_\Omega Z_\chi(p_\theta)dx.$$

Historically, the notion of escort distributions was introduced by Beck and Schlögl [4]. Later, Naudts introduced more generalized framework in [17]. The author recently introduced a sequential structure of escort distributions [12].

5. Geometry of deformed exponential families

In this section, we consider the geometry of deformed exponential families. In particular, we review the results in [12] and [15]. Since S_χ has a sequential structure of expectations, we can define a sequential structure of statistical manifolds.

For a χ-exponential family S_χ, we define a $(0,2)$-tensor field $g^{(n)}$ by

$$g^{(n)}_{ij}(\theta) := \int_\Omega (\partial_i \ln_\chi p_\theta)(\partial_j \ln_\chi p_\theta)P_{\chi,(n)}(x;\theta)dx.$$

We assume that $g^{(n)}$ is a Riemannian metric on S_χ for all $n \in \mathbf{N}$. Similarly, we define a $(0,3)$-tensor field $C^{(n)}$ by

$$C_{ijk}^{(n)}(\theta) := \int_\Omega (\partial_i \ln_\chi p_\theta)(\partial_j \ln_\chi p_\theta)(\partial_k \ln_\chi p_\theta) P_{\chi,(n+1)}(x;\theta)dx.$$

The tensor field $C^{(n)}$ is regarded as a cubic form on a statistical manifold. As a consequence, we obtain a sequence of statistical manifolds

$$(S_\chi, g^{(1)}, C^{(1)}) \;\to\; (S_\chi, g^{(2)}, C^{(2)}) \;\to\; \cdots \to (S_\chi, g^{(n)}, C^{(n)}) \;\to\; \cdots.$$

5.1. Statistical manifold structure derived from the first escort expectation

Let us consider geometric structures for S_χ derived from the first escort expectation. We define a Riemannian metric g^M on S_χ by

$$g_{ij}^M(\theta) := \int_\Omega (\partial_i \ln_\chi p(x;\theta))\,(\partial_j p(x;\theta))\,dx. \tag{12}$$

The metric g^M is a generalization of g^F in the representation (2).

We define two affine connections $\nabla^{M(e)}$ and $\nabla^{M(m)}$ by

$$\Gamma_{ij,k}^{M(e)}(\theta) = \int_\Omega (\partial_i \partial_j \ln_\chi p(x;\theta))(\partial_k p(x;\theta))dx, \tag{13}$$

$$\Gamma_{ij,k}^{M(m)}(\theta) = \int_\Omega (\partial_k \ln_\chi p(x;\theta))(\partial_i \partial_j p(x;\theta))dx, \tag{14}$$

respectively. By differentiating Equation (12), we can easily find that $\nabla^{M(e)}$ and $\nabla^{M(m)}$ are mutually dual with respect to g^M. The cubic form C^M for $(S_\chi, \nabla^{M(e)}, g^M)$ is given by

$$C^M(X,Y,Z) := g^M(\nabla_X^{M(m)}Y, Z) - g^M(\nabla_X^{M(e)}Y, Z).$$

The triplet (S_χ, g^M, C^M) then becomes a statistical manifold.

Theorem 5.1 ([15]). *For a χ-exponential family S_χ, the statistical manifold $(S_\chi, g^{(1)}, C^{(1)})$ coincides with (S_χ, g^M, C^M).*

Proof. From the definitions of χ-logarithm and the escort distribution, using the chain rule of differentiation, we obtain

$$\partial_i p_\theta = \frac{\partial_i p_\theta}{\chi(p_\theta)}\chi(p_\theta) = (\partial_i \ln_\chi p_\theta)P_{\chi,(1)}(x;\theta), \tag{15}$$

$$\partial_i P_{\chi,(1)}(x;\theta) = (\partial_i \ln_\chi p_\theta)P_{\chi,(2)}(x;\theta). \tag{16}$$

From Equations (15) and (16), we have

$$\partial_i \partial_j p_\theta = (\partial_i \partial_j \ln_\chi p_\theta) P_{\chi,(1)}(x;\theta) + (\partial_i \ln_\chi p_\theta)(\partial_i \ln_\chi p_\theta) P_{\chi,(2)}(x;\theta). \quad (17)$$

Therefore, we obtain

$$g_{ij}^M(\theta) = \int_\Omega (\partial_i \ln_\chi p_\theta)(\partial_j p_\theta) dx$$
$$= \int_\Omega (\partial_i \ln_\chi p_\theta)(\partial_j \ln_\chi p_\theta) P_{\chi,(1)}(x;\theta) dx$$
$$= g^{(1)}(\theta).$$

From the definition of deformed exponential family and Equation (13), $\Gamma_{ij,k}^{M(e)}$ always vanishes. Therefore, from Equations (15) and (17), we obtain

$$C_{ijk}^M(\theta) = \Gamma_{ij,k}^{M(m)}(\theta) = \int_\Omega (\partial_k \ln_\chi p_\theta)(\partial_i \partial_j p_\theta) dx$$
$$= \int_\Omega (\partial_k \ln_\chi p_\theta)\{(\partial_i \partial_j \ln_\chi p_\theta) P_{\chi,(1)}(x;\theta)$$
$$+ (\partial_i \ln_\chi p_\theta)(\partial_i \ln_\chi p_\theta) P_{\chi,(2)}(x;\theta)\} dx$$
$$= \int_\Omega (\partial_i \partial_j \ln_\chi p_\theta)(\partial_k p_\theta) dx$$
$$+ \int_\Omega (\partial_k \ln_\chi p_\theta)(\partial_j \ln_\chi p_\theta)(\partial_i \ln_\chi p_\theta) P_{\chi,(2)}(x;\theta) dx$$
$$= C_{ijk}^{(1)}(\theta),$$

and get the conclusion.   □

Since $\nabla^{M(e)}$ is a flat connection, the quadruplet $(S_\chi, g^M, \nabla^{M(e)}, \nabla^{M(m)})$ is a dually flat space. Set the integral of deformed exponential function by

$$U_\chi(s) := \exp_\chi^{(-1)}(s) := \int_0^s (\exp_\chi t)\, dt.$$

Then the canonical divergence for $(S_\chi, g^M, \nabla^{M(e)}, \nabla^{M(m)})$ is given by

$$D_\chi(p,r) = \int_\Omega \{U_\chi(\ln_\chi r(x)) - U_\chi(\ln_\chi p(x)) - p(x)(\ln_\chi r(x) - \ln_\chi p(x))\} dx,$$

where $p = p(x) = p(x;\theta)$ and $r = r(x) = p(x;\theta')$. In information geometry, $D_\chi(r,p)$ is known as a U-divergence [16, 19]. In the case of q-exponential family, the divergence D_χ coincides with a β-divergence [13].

5.2. Statistical manifold structure derived from the second escort expectation

Next, we define another statistical manifold structure for S_χ. Since we assumed that the normalization ψ is strictly convex, we can define a Riemannian metric g^χ and a cubic form C^χ by

$$g_{ij}^\chi(\theta) := \partial_i \partial_j \psi(\theta),$$
$$C_{ijk}^\chi(\theta) := \partial_i \partial_j \partial_k \psi(\theta).$$

The metric g^χ is called a χ-*Fisher metric*, and C^χ is called a χ-*cubic form* (cf. [3]). Obviously, the triplet (S_χ, g^χ, C^χ) is a statistical manifold. From Equation (4), we can define an affine connection $\nabla^{\chi(\alpha)}$ by

$$g^\chi(\nabla_X^{\chi(\alpha)} Y, Z) := g^\chi(\nabla_X^{\chi(0)} Y, Z) - \frac{\alpha}{2} C^\chi(X, Y, Z),$$

where $\nabla^{\chi(0)}$ is the Levi-Civita connection with respect to g^χ. By standard arguments in Hessian geometry [21], the quadruplet $(S_\chi, g^\chi, \nabla^{\chi(1)}, \nabla^{\chi(-1)})$ is a dually flat space.

Theorem 5.2 ([15]). *For a χ-exponential family S_χ, the second statistical manifold structure $(S_\chi, g^{(2)}, C^{(2)})$ and (S_χ, g^χ, C^χ) have the following relations.*

$$g_{ij}^{(2)}(x; \theta) = Z_\chi(p_\theta) g_{ij}^\chi(\theta), \tag{18}$$
$$C_{ijk}^{(2)}(x; \theta) = Z_\chi(p_\theta) C_{ijk}^\chi(\theta)$$
$$+ g_{ij}^\chi(\theta) \partial_k Z_\chi(p_\theta) + g_{jk}^\chi(\theta) \partial_i Z_\chi(p_\theta) + g_{ki}^\chi(\theta) \partial_j Z_\chi(p_\theta), \tag{19}$$

where $Z_\chi(p_\theta)$ is the normalization for the escort distribution $P_{\chi,(1)}(x; \theta)$.

Proof. Let us show the relations of Riemannian metrics $g^{(2)}$ and g^χ.
By differentiating $p(x; \theta) \in S_\chi$ twice, from Equation (17), we obtain

$$\partial_i \partial_j p(x; \theta) = P_{\chi,(2)}(x; \theta)(\partial_i \ln_\chi p_\theta)(\partial_j \ln_\chi p_\theta) - P_{\chi,(1)}(x; \theta)\partial_i \partial_j \psi(\theta). \tag{20}$$

Since we assumed that an integration and a differentiation are interchangeable, we can fined that $\int_\Omega \partial_i \partial_j p(x; \theta) dx = 0$. By integrating Equation (20), we obtain (18).

We can show that the relation between $C^{(2)}$ and C^χ from the third derivatives. □

We remark that the statistical manifold $(S_\chi, g^{(2)}, C^{(2)})$ cannot determine a dually flat structure in general whereas (S_χ, g^χ, C^χ) determines a dually flat structure. To describe the relation of two statistical manifold structures, we need an argument of a generalized conformal equivalence relation for statistical manifolds. (See [9] and [11].)

We define a generalization of KL-divergence by

$$D^\chi(p,r) := E^{esc}_{\chi,p}[\ln_\chi p(x) - \ln_\chi r(x)].$$

The canonical divergence $D(p,r)$ for $(S_\chi, g^\chi, \nabla^{\chi(1)}, \nabla^{\chi(-1)})$ coincides with $D^\chi(r,p)$. In the case of q-exponential family, the divergence

$$D^T(p,q) := E^{esc}_{q,p}[\ln_q p(x) - \ln_q r(x)]$$

is known as a *normalized Tsallis relative entropy*.

6. q-maximum likelihood estimator for multivariate Student's t-distributions

In this section, we consider a generalization of maximum likelihood method for deformed exponential families. In particular, we give a q-maximum likelihood estimator for the set of multivariate Student's t-distributions.

Let us recall the notion of independence of random variables. Suppose that two random variables X and Y are distributed according to probability distributions $p_1(x)$ and $p_2(y)$, respectively. We say that X and Y are *independent* if $p_1(x)$ and $p_2(y)$ are the marginal distributions of the joint probability distribution $p(x,y)$. That is,

$$p(x,y) = p_1(x)p_2(y), \quad (\forall x \in \Omega_x, \ \forall y \in \Omega_y), \tag{21}$$

where Ω_x and Ω_y are the total sample spaces for X and Y, respectively. If $p_1(x) > 0$ ($\forall x \in \Omega_x$) and $p_2(y) > 0$ ($\forall y \in \Omega_y$), the formula (21) can be written by

$$p(x,y) = \exp[\ln p_1(x) + \ln p_2(y)].$$

Hence the notion of independence is attributed to the duality of the exponential and the logarithm. We can generalize the notion of independence by using a deformed exponential and a deformed logarithm functions.

Suppose that $x > 0, y > 0$ and $x^{1-q} + y^{1-q} - 1 > 0$ ($q > 0$). The q-product of x and y is defined by

$$x \otimes_q y := \left[x^{1-q} + y^{1-q} - 1\right]^{1/(1-q)} = \exp_q[\ln_q x + \ln_q y].$$

A generalization of the law of exponents holds under the q-product, that is

$$\exp_q(x+y) = \exp_q x \otimes_q \exp_q y,$$
$$\ln_q(x \otimes_q y) = \ln_q x + \ln_q y.$$

Suppose that X_i is a random variable on Ω_i distributed according to $p_i(x)$ $(i = 1, 2, \ldots, N)$. We define a positive valued function f by

$$f(x_1, x_2, \ldots, x_N) := p_1(x_1) \otimes_q p_2(x_2) \otimes_q \cdots \otimes_q p_N(x_N),$$

and we say that $X_1, X_2, \ldots, X_N$ are q-independent. We remark that the joint distribution $f(x_1, x_2, \ldots, x_N)$ is not a probability distribution in general. In addition, the support of this function does not coincide with the entire product space $\Omega_1 \times \Omega_2 \times \cdots \Omega_N$. We denote the support of this function f by $\Omega^N = \text{Supp}\{p(x_1, x_2, \ldots, x_N)\} \subset \Omega_1 \times \Omega_2 \times \cdots \Omega_N$.

Let us consider normalizations to obtain a probability distribution. We say that $X_1, X_2, \ldots, X_N$ are q-independent with m-normalization if

$$p(x_1, x_2, \ldots, x_N) = \frac{f(x_1, x_2, \ldots, x_N)}{Z_{p_1, p_2, \cdots, p_N}}$$
$$= \frac{p_1(x_1) \otimes_q p_2(x_2) \otimes_q \cdots \otimes_q p_N(x_N)}{Z_{p_1, p_2, \cdots, p_N}},$$

where $Z_{p_1, p_2, \cdots, p_N}$ is the normalization of $f(x_1, x_2, \ldots, x_N)$ defined by

$$Z_{p_1, p_2, \cdots, p_N} := \int \cdots \int_{\Omega^N} p_1(x_1) \otimes_q p_2(x_2) \otimes_q \cdots \otimes_q p_N(x_N) dx_1 \cdots dx_N.$$

Similarly, we say that $X_1, X_2, \ldots, X_N$ are q-independent with e-normalization if

$$p(x_1, x_2, \ldots, x_N) = p_1(x_1) \otimes_q p_2(x_2) \otimes_q \cdots \otimes_q p_N(x_N) \otimes_q (-c),$$

where c is the normalization defined by

$$\int \cdots \int_{\Omega^N} p_1(x_1) \otimes_q p_2(x_2) \otimes_q \cdots \otimes_q p_N(x_N) \otimes_q (-c) dx_1 \cdots dx_N = 1.$$

Example 6.1. Suppose that X_1 and X_2 are random variables distributed according to univariate Student's t-distributions (or q-normal distributions) $p_1(x_1)$ and $p_2(x_2)$, respectively, with same parameter q $(1 < q < 2)$. That is,

$$p_i(x_i; \mu_i, \sigma_i) = \frac{\Gamma\left(\frac{1}{q-1}\right)}{\sqrt{\pi} \sqrt{\frac{3-q}{q-1}} \Gamma\left(\frac{3-q}{2(q-1)}\right) \sigma_i} \left[1 - (1-q)\frac{(x_i - \mu_i)^2}{(3-q)\sigma_i^2}\right]^{1/(1-q)},$$

for $i = 1, 2$, where $\mu_i \in (-\infty, \infty)$ is a location parameter and $\sigma_i \in (0, \infty)$ is a scale parameter. Then there exists a bivariate Student's t-distribution $p(x_1, x_2)$ such that X_1 and X_2 are q-independent with e-normalization.

For further details, see [20]. We remark that, even if X_1 and X_2 are independent, the joint probability $p_1(x_1)p_2(x_2)$ is not a bivariate Student's t-distribution [7].

Let $S_q = \{p(x;\theta) \mid \theta \in \Theta\}$ be a q-exponential family. Suppose that $x_1, \ldots, x_N$ are N-observations from $p(x;\theta) \in S_q$. We define a q-*likelihood function* $L_q(\theta)$ by

$$L_q(\theta) = p(x_1;\theta) \otimes_q p(x_2;\theta) \otimes_q \cdots \otimes_q p(x_N;\theta).$$

By taking a q-logarithm, we set

$$l_q(\theta) := \ln_q L_q(\theta) = \sum_{i=1}^{N} \ln_q p(x_i;\theta),$$

and call it the q-*logarithm q-likelihood function* (or q-*log likelihood* for short). The parameter which maximizes the q-likelihood function is called the q-*maximum likelihood estimator* (q-MLE), that is,

$$\hat{\theta} = \arg\max_{\theta \in \Theta} L_q(\theta) \quad \left(= \arg\max_{\theta \in \Theta} l_q(\theta) \right).$$

Let us consider geometry of maximum q-likelihood estimators. Suppose that M is a submanifold in S_q and that $x_1, \ldots, x_N$ are N-observations generated from $p(x;\theta(u)) \in M \subset S_q$. The q-log likelihood function is given by

$$\ln_q L_q(\theta(u)) = \sum_{j=1}^{N} \ln_q p(x_j;\theta(u)) = \sum_{j=1}^{N} \left\{ \sum_{i=1}^{n} \theta^i(u) F_i(x_j) - \psi(\theta(u)) \right\}$$

$$= \sum_{i=1}^{n} \theta^i(u) \sum_{j=1}^{N} F_i(x_j) - N\psi(\theta(u)).$$

Therefore, the geometric interpretation of q-MLE for S_q is quite similar to MLE for an exponential family S_e. In fact, the normalized Tsallis relative entropy for S_q can be given by

$$D_q^T(p(\hat{\eta}), p(\theta(u))) = \phi(\hat{\eta}) - \frac{1}{N}\ln_q L_q(\theta(u)).$$

This implies that the q-likelihood attains the maximum if and only if the normalized Tsallis relative entropy from $p(\hat{\eta}) \in S_q$ to M attains the minimum.

Acknowledgments

The author would like to express his sincere gratitude to the referee for giving insightful comments to improve this paper. This work was partially supported by JSPS KAKENHI No. JP26108003, JP15K04842 and JP16KT0132.

References

[1] S. Amari, *Information Geometry and Its Applications*; Springer, Tokyo, 2016.

[2] S. Amari and H. Nagaoka, *Method of Information Geometry*, Amer. Math. Soc., Providence, Oxford University Press, Oxford, 2000.

[3] S. Amari, A. Ohara and H. Matsuzoe, Geometry of deformed exponential families: invariant, dually-flat and conformal geometry, *Physica A.*, **391**, 4308-4319, (2012).

[4] C. Beck and F. Schlögl, *Thermodynamics of Chaotic Systems: An Introduction*; Cambridge University Press, Cambridge, UK, 1993.

[5] S. Eguchi, Geometry of minimum contrast, *Hiroshima Math. J.*, **22**, 631–647, (1992).

[6] G. Kaniadakis, Theoretical foundations and mathematical formalism of the power-law tailed statistical distributions. *Entropy*, **15**, 3983–4010, (2013).

[7] S. Kotz and S. Nadarajah, *Multivariate t Distributions and Their Applications*, Cambridge University Press, 2004.

[8] T. Kurose, On the divergences of 1-conformally flat statistical manifolds, *Tôhoku Math. J.*, **46**, 427–433, (1994).

[9] T. Kurose, Conformal-projective geometry of statistical manifolds, *Interdiscip. Inform. Sci.*, **8**, 89–100, (2002).

[10] S.L. Lauritzen, Statistical manifolds, *Differential geometry in statistical inferences*, IMS Lecture Notes Monograph Series 10, Institute of Mathematical Statistics, Hayward California, 96–163, (1987).

[11] H. Matsuzoe, Geometry of contrast functions and conformal geometry, *Hiroshima Math. J.*, **29**, 175 – 191, (1999).

[12] H. Matsuzoe, A sequence of escort distributions and generalizations of expectations on q-exponential family, *Entropy*, **19**, no. 1, 7, (2017).

[13] H. Matsuzoe and M. Henmi, Hessian structures and divergence functions on deformed exponential families. *Geometric Theory of Information, Signals and Communication Technology*, Springer: Basel, Switzerland, 57–80, (2014).

[14] H. Matsuzoe and A. Ohara, Geometry for q-exponential families, in *Recent progress in Differential Geometry and its Related Fields*, T. Adachi, H. Hashimoto and M. Hristov eds, World Sci. Publ., New Jersey, (2011), 55-71.

[15] H. Matsuzoe, A. Scarfone and T. Wada, A sequential structure of statistical manifolds on deformed exponential family, to appear in *Lecture Notes in Comp. Sci.*.

[16] N. Murata, T. Takenouchi, T. Kanamori and S. Eguchi, Information geometry of U-boost and Bregman divergence, *Neural Comput.*, **16**, 1437-1481, (2004).

[17] J. Naudts, Estimators, escort probabilities, and ϕ-exponential families in statistical physics, *J. Inequal. Pure Appl. Math.*, **5**, no. 102, (2004).

[18] J. Naudts, *Generalised Thermostatistics*, Springer-Verlag, 2011.

[19] A. Ohara and T. Wada, Information geometry of q-Gaussian densities and behaviors of solutions to related diffusion equations, *J. Phys. A: Math. Theor.*, **43** No.035002, (2010).

[20] M. Sakamoto and H. Matsuzoe, A generalization of independence and multivariate Student's t-distributions, *Lecture Notes in Comp. Sci.*, Springer, **9389**, 740–749, (2015).

[21] H. Shima, *The Geometry of Hessian Structures*, World Scientific, 2007.

[22] C. Tsallis, *Introduction to Nonextensive Statistical Mechanics: Approaching a Complex World*, Springer, New York, 2009.

[23] J. Zhang, On monotone embedding in information geometry, *Entropy*, **17**, 4485–4499, (2015).

Received July 24, 2017
Revised   July 29, 2017

Printed in the United States
By Bookmasters